W9-BSK-148

1995 YEARBOOK

Connecting Mathematics across the Curriculum

Peggy A. House

1995 Yearbook Editor
Northern Michigan University

Arthur F. Coxford

General Yearbook Editor
University of Michigan

NATIONAL COUNCIL OF TEACHERS OF MATHEMATICS

Library of Congress Cataloging-in-Publication Data:

Connecting mathematics across the curriculum / [edited by] Peggy A. House, Arthur F. Coxford.
p. cm. — (Yearbook ; 1995)
Includes bibliographical references.
ISBN 0-87353-394-1
1. Mathematics—Study and teaching. I. House, Peggy. II. Coxford, Arthur F. III. Series: Yearbook (National Council of Teachers of Mathematics) ; 1995.
QA1.N3 1995
[QA11]
510'.71'2 s—dc20
[510'.71'2] 94-48261
CIP

The publications of the National Council of Teachers of Mathematics present a variety of viewpoints. The views expressed or implied in this publication, unless otherwise noted, should not be interpreted as official positions of the Council.

Printed in the United States of America

Contents

Preface vii

PART 1: GENERAL ISSUES

1. The Case for Connections 3
Arthur F. Coxford
University of Michigan
Ann Arbor, Michigan

2. Connections as Problem-Solving Tools 13
Theodore R. Hodgson
Montana State University
Bozeman, Montana

3. Connecting School Science and Mathematics 22
Donna F. Berlin
National Center for
Science Teaching and Learning
Columbus, Ohio
Arthur L. White
Ohio State University
Columbus, Ohio

4. Using Ethnomathematics to Find Multicultural Mathematical Connections 34
Lawrence Shirley
Towson State University
Baltimore, Maryland

PART 2: CONNECTIONS WITHIN MATHEMATICS

5. Connecting Number and Geometry 45
Lowell Leake
University of Cincinnati
Cincinnati, Ohio

6. Using Functions to Make Mathematical Connections 54
Roger P. Day
Illinois State University
Normal, Illinois

7. Making Connections with Transformations in Grades K–8 65
Rheta N. Rubenstein
University of Windsor
Windsor, Ontario
Denisse R. Thompson
University of South Florida
Tampa, Florida

8. Transformations: Making Connections in High School Mathematics 79
Mary L. Crowley
Dalhousie University
Halifax, Nova Scotia

9. Using Transformations to Foster Connections 92
Daniel B. Hirschhorn
Illinois State University
Normal, Illinois
Steven S. Viktora
New Trier High School
Winnetka, Illinois

10. Connecting Mathematics with Its History: A Powerful, Practical Linkage 104
Luetta Reimer
Fresno Pacific College
Fresno, California
Wilbert Reimer
Fresno Pacific College
Fresno, California

PART 3: CONNECTIONS ACROSS THE ELEMENTARY SCHOOL CURRICULUM

11. Learning Mathematics in Meaningful Contexts: An Action-Based Approach in the Primary Grades 116
Sydney L. Schwartz
Queens College
City University of New York
Flushing, New York
Frances R. Curcio
Queens College
City University of New York
Flushing, New York

12. Measurement in a Primary-Grade Integrated Curriculum 124
Lynn Rhone
Arrowhead Elementary School
Aurora, Colorado

13. Connecting Literature and Mathematics 134
David J. Whitin
University of South Carolina
Columbia, South Carolina

14. Connecting Reasoning and Writing in Student "How to" Manuals 142

Neal F. Grandgenett
University of Nebraska at Omaha
Omaha, Nebraska

John W. Hill
University of Nebraska at Omaha
Omaha, Nebraska

Carol V. Lloyd
University of Nebraska at Omaha
Omaha, Nebraska

15. Connecting Mathematics and Physical Education through Spatial Awareness 147

Diana V. Lambdin
Indiana University
Bloomington, Indiana

Dolly Lambdin
University of Texas at Austin
Austin, Texas

PART 4: CONNECTIONS ACROSS THE MIDDLE SCHOOL CURRICULUM

16. Seeing and Thinking Mathematically in the Middle School 153

Glenn M. Kleiman
Educational Development Center
Newton, Massachusetts

17. Projects in the Middle School Mathematics Curriculum 159

Stephen Krulik
Temple University
Philadelphia, Pennsylvania

Jesse Rudnick
Temple University
Philadelphia, Pennsylvania

18. Carpet Laying: An Illustration of Everyday Mathematics 163

Joanna O. Masingila
Syracuse University
Syracuse, New York

19. Mathematics and Quilting 170

Kathryn T. Ernie
University of Wisconsin at River Falls
River Falls, Wisconsin

20. Randomness: A Connection to Reality 177
Donald J. Dessart
University of Tennessee
Knoxville, Tennessee

PART 5: CONNECTIONS ACROSS THE HIGH SCHOOL CURRICULUM

21. Connecting Geometry with the Rest of Mathematics 183
Albert A. Cuoco
Education Development Center
Newton, Massachusetts
E. Paul Goldenberg
Education Development Center
Newton, Massachusetts
June Mark
Education Development Center
Newton, Massachusetts

22. Forging Links with Projects in Mathematics 198
John W. McConnell
Glenbrook South High School
Glenview, Illinois

23. Baseball Cards, Collecting, and Mathematics 210
Vincent P. Schielack, Jr.
Texas A&M University
College Station, Texas

24. Experiencing Functional Relationships with a Viewing Tube 219
Melvin R. (Skip) Wilson
University of Michigan
Ann Arbor, Michigan
Barry E. Shealy
SUNY Buffalo
Buffalo, New York

25. Breathing Life into Mathematics 225
Kristine Malia Johnson
LaSalle Institute
Albany, New York
Carolyn Leigh Litynski
LaSalle Institute
Albany, New York

26. Students' Reasoning and Mathematical Connections in the Japanese Classroom 233
Keiko Ito-Hino
Tokyo, Japan

Preface

One of the four cornerstones of the NCTM *Curriculum and Evaluation Standards for School Mathematics* asserts that connecting mathematics to other mathematics, to other subjects of the curriculum, and to the everyday world is an important goal of school mathematics. Among recent reports calling for reform in mathematics education, there is widespread consensus that mathematics must be made accessible to all students, that it must be presented as a connected discipline rather than a set of discrete topics, and that it must be learned in meaningful contexts that connect mathematics to other subjects and to the interests and experience of students.

This yearbook illustrates the connections and uses of mathematics within mathematics itself, between mathematics and other disciplines, and in the life, culture, and occupational experiences of adult communities. It is designed to help classroom teachers, teacher educators, supervisors, and curriculum developers broaden their views of mathematics and to suggest practical strategies for engaging students in exploring the connectedness of mathematics.

The yearbook is organized in five parts. Part One, chapters 1 through 4, addresses general issues and various perspectives as they relate to the development and uses of mathematical connections. The papers in this section explore the meaning and scope of mathematical connections and the role of connections in teaching and learning mathematics.

An important outcome of mathematics education is to present mathematics as a unified discipline, a woven fabric rather than a patchwork of discrete topics. Papers in Part Two, chapters 5 through 10, focus on connections within mathematics itself. They illustrate how concepts emerge in the early grades and grow in sophistication and applicability throughout the mathematics curriculum. Papers in this section illustrate how numerical concepts and the basic operations of arithmetic are connected to geometric concepts and how the all-important concept of *function* develops across the curriculum. One important conceptual strand, that of *transformations,* is developed in detail through three related papers as an illustration of the richness of mathematical topics.

But mathematics is not an isolated body of knowledge. To be useful, mathematics should be taught in contexts that are meaningful and relevant to learners. Papers in Parts Three, Four, and Five illustrate opportunities to connect mathematics across the curriculum of the elementary, middle, and secondary school years, respectively. The examples include connections to other school subjects and to mathematics as it is used in adult life.

The production of this yearbook represents the efforts of many dedicated professionals over a three-year period. More than sixty manuscripts were received in response to our call for papers; fewer than half of those submitted could be accepted. To all who gave of their time and talents to answer the call for papers, and especially to the thirty-nine authors of the twenty-six papers in this publication, the Editorial Panel and I express our deep appreciation.

The tasks of developing the guidelines for this yearbook, of reviewing all the submitted papers, of selecting the papers to be included, and of guiding the development of the final product fell to the Editorial Panel, a talented and generous group of colleagues for whom I have great admiration and sincere gratitude:

Jerry Becker	Southern Illinois University
Gary Froelich	COMAP (Consortium for Mathematics and Its Applications)
Linda Sheffield	Northern Kentucky University
Irvin Vance	Michigan State University
Arthur Coxford	University of Michigan

A special word of thanks is owed to Art Coxford, series editor for the 1993 through 1995 Yearbooks, who was a full and active participant in the workings of the Editorial Panel and whose support and wise counsel were never lacking. Our thanks, as well, go to the talented editorial and production staff at the NCTM Headquarters Office, whose efforts transformed a collection of papers into a finished publication.

The writers of these papers have endeavored to make mathematics come alive for the readers of this yearbook. In so doing, they present us all with a challenge to do likewise for the students we teach. It is our hope that the ideas brought forth in this publication will become catalysts for teachers to develop and implement many more rich contexts for connecting students to the fascinating world of mathematics.

PEGGY A. HOUSE
1995 Yearbook Editor

PART 1

General Issues

1

The Case for Connections

Arthur F. Coxford

THE desire to emphasize the connections in mathematics, to foster mathematical thinking in other disciplines, and to contextualize mathematics so that learners will see mathematics as a means to help make sense of their world is not a phenomenon unique to the late twentieth century. The National Committee on Mathematical Requirements (1923) recommended an integrated curriculum emphasizing functionality as early as 1923. The Commission on the Secondary School Curriculum of the Progressive Education Association, in 1940, also emphasized a connected curriculum (pp. 44–46):

> Adolescents encounter certain problems as they strive to meet their needs in the basic aspects of living.... [Mathematics makes its] special contribution whenever quantitative data ... and relationships of space and form are encountered. The highly effective special symbolism and methods of mathematics have been developed in order to treat just such aspects of experience. ... The teacher of mathematics bears the responsibility of equipping students to solve problems with the aid of mathematical concepts and methods as they seek to meet their needs throughout life. In this process he also has the responsibility of throwing light on the nature of problem solving.

The most recent expression of the desire for connections in mathematics occurs in the *Curriculum and Evaluation Standards for School Mathematics* (National Council of Teachers of Mathematics [NCTM] 1989). One of the four common standards at each grade level calls for opportunities for students to experience the connections and interplay of various mathematical topics. How students are to experience the connections and interplay of mathematical topics is not easily summarized in a few words. It has many facets, each of which deserves a comprehensive discussion of its own. The standards themselves are broad in their characterization of the ways students can experience mathematical connections and as a result—

- link conceptual and procedural knowledge;
- use mathematics in other curriculum areas;

The writer thanks the members of the Editorial Panel, Jerry Becker, Gary Froelich, Linda Sheffield, Irv Vance, and Peggy House, for their assistance in developing this paper.

- use mathematics in daily life activities;
- see mathematics as an integrated whole;
- apply mathematical thinking and modeling to solve problems that arise in other disciplines, such as art, music, psychology, science, and business;
- use and value the connections among mathematical topics;
- recognize equivalent representations of the same concept.

For this article, the concept of mathematical connections is assumed to have three related aspects: (1) unifying themes, (2) mathematical processes, and (3) mathematical connectors. No claim regarding the completeness or independence of these aspects is implied. Rather, the three aspects are used to organize the examples, illustrations, suggestions, and discussions.

UNIFYING THEMES

The reader may think of many possible themes, such as modeling, application, chance, and number, that may be used to draw attention to the connected nature of mathematics. To illustrate the power and pervasiveness of themes, we chose the following three for discussion: (1) change, (2) data, and (3) shape.

Change

Change is pervasive in life and in mathematics. In the elementary school, children can be asked to look for, and investigate, changes in results when an addend, factor, or divisor is changed by 1, by 2, or by some other amount. They can explore the changes that occur in a geometric shape when it is transformed somehow, perhaps by turning, sliding, or flipping it. Or, if models like those one can make with straws and pipe cleaners are available, they can explore which polygons and polyhedra change shape and which are rigid when force is applied at a vertex. Exploring what remains unchanged is also interesting. Computer software that allows the user to change shapes dynamically furnishes another avenue for students to experience change.

At more advanced levels, the notion of change may help connect algebra, geometry, discrete mathematics, and calculus. For example, how is a constant rate of change related to lines and linear equations? What changes occur in the graph of a function when a coefficient in the equation of the function is changed? Or, what change in the value of a function accompanies a change in the independent variable? Are the patterns of change related? Does a change in y depend on the magnitude of the value of x before it is changed? How does the perimeter or area of a plane shape change when it is transformed using isometries, size transformations, shears, or some unspecified linear transformation? What happens to a route matrix for a mail delivery system when changes occur in the number and locations of delivery substations? What is the instantaneous rate of change of a function at x_0?

Each of these questions suggests opportunities to connect mathematical topics by relating them through the theme of change. The list is not exhaustive; rather, it is meant to be suggestive of how the theme of change can be used to emphasize the connected nature of the mathematics.

Data

Data is another theme that furnishes many opportunities to emphasize connections in all aspects of mathematics. Elementary school activities in which students collect information about a concept, a procedure, or a problem situation are examples of connecting through data. Measuring the height of each student to the nearest centimeter, displaying the data graphically and asking students to describe what they see (e.g., What is the most common height? What is a typical height? Who is tallest or shortest?) are examples.

Whenever we seek to generalize inductively, data are gathered and analyzed to support or refute the generalization. For example, data can easily be gathered to check conjectures about the nature of the sum (or product) of two even or two odd numbers. Using measurements of their bodies, such as height, weight, arm span, the circumference of the head, and so on, children can investigate possible relations among the measures. For pairs of data, a graphical representation can be especially informative.

Perhaps one of the most powerful ways in which data connect mathematics is that they provide *real* contexts and purposes for studying a specific topic. Linearity is a prime example. Bivariate data, in the form of scatterplots, provide context and motivation for learning about linear functions, since the data may often be summarized by a linear function. Students can then see some purpose in finding values of y for a given value of x: it is the way the linear model is used to predict. Also, the importance of the slope as a measure of the rate of change of an entire data set is evident. Thus, data provide a means to help students see mathematics as a sense-making tool related to their world.

It is also easy to emphasize data while studying other topics. For example, a class project of measuring the angles of polygons can lead to a data set representing the number of sides and the sum of the angle measures. Students, being human, will generate data that will show some variation. The data can be plotted, a line fitted, and an equation generated as shown in figure 1.1. This equation can be used to develop the actual formula, $S = (n - 2)180$, or it can be developed using other means, such as dissecting polygons into triangles. The two approaches can be compared to illustrate both the power of data-analysis techniques in generating good mathematical models and the need for more formal mathematical procedures to derive the model using the characteristics of the mathematical system itself.

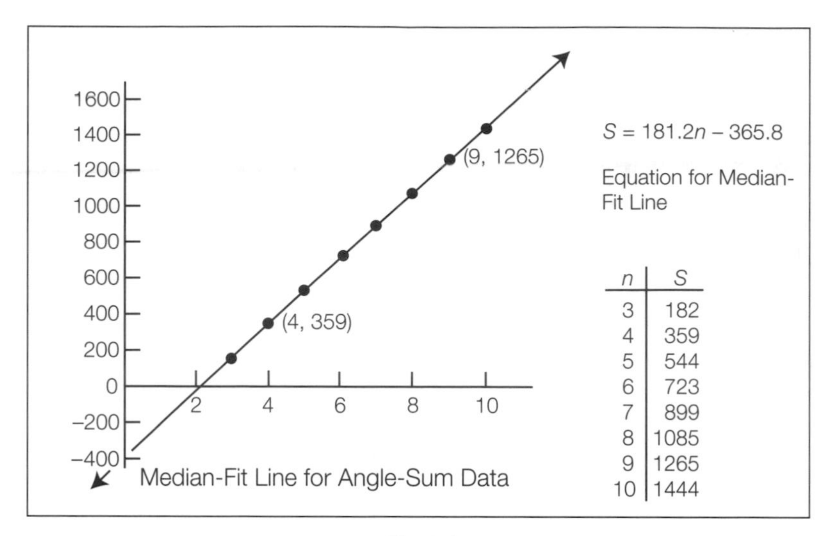

n	S
3	182
4	359
5	544
6	723
7	899
8	1085
9	1265
10	1444

Fig. 1.1

Shape

Shape is another unifying theme that can be used to emphasize connections in mathematics. The notion of shape is central to the study of geometry, where two- and three-dimensional shapes and their properties and relations are the primary focus. In the elementary school, students learn to recognize, name, and work with shapes. Early in the elementary school, students check models for folding symmetry—for example, in February when teachers and children make hearts for Valentine's Day. This is mathematics at work. Later, the areas of plane shapes and volumes of space shapes connect the mathematics to the student's world and to real-world problem solving. Numbers and shape are also related through figurate numbers, such as triangular and square numbers. For example, triangular numbers, 1, 3, 6, 10, 15, …, can be placed into a triangular configuration. Similarly, square numbers, 1, 4, 9, 16, …, can be placed in square arrays.

Shape comes into play throughout the school curriculum in other ways also. For example, in work on data representation and analysis, a scatterplot may have a shape that suggests whether it should be modeled by a linear, exponential, or power function, and it is the shape of the plot that suggests which analytic tool to try. A distribution of data represented by a histogram may be either roughly symmetrical or skewed one way or the other. Shape is stressed when students are asked to discuss what the shape of the distribution suggests about the variable. Similarly, in the study of functions, certain families of curves have shapes that are characteristic and that help to identify the curve without seeing an equation. Shape also furnishes clues about the nature of the relations that produced the graph.

The use of shape here is heuristic. That is, it is an aid to solving a problem or identifying an underlying pattern. It helps the mathematics retain its sense, but it does not abrogate the need for establishing truth using accepted mathematical procedures.

In the preceding discussion, we have tried to suggest how certain unifying themes can be (and should be) used to connect the mathematics to itself, to the larger world, and to the learner. More fully developed examples appear in other articles in this yearbook. As you work to improve your practice, think about unifying themes, these or others, that learners can use to connect the mathematics being studied internally, externally, and personally.

MATHEMATICAL PROCESSES

The *mathematical processes* aspect of mathematical connections includes (1) *representation,* (2) *application,* (3) *problem solving,* and (4) *reasoning.* These four categories of activity should continue in all the mathematical work done by students from kindergarten through independent learning as an adult. At each stage, the specific activities should be developmentally appropriate, but none should be slighted or omitted.

Representation

Representation is an important mathematical process in the elementary school curriculum. The usual form and sequence it takes begins with concrete representations created by teachers and by students, then proceeds to pictorial and abstract representations for such topics as addition, subtraction, multiplication, division, whole numbers, fractions, place value, and decimals. For example, teachers may begin representing small whole numbers with real objects, such as toys, books, or beans. Students learn to recognize the concept of number in such settings, to produce concrete representations in response to oral commands, and to apply the oral names to pictorial and symbolic representations. For students to have a deep understanding of a concept, they need to make the connections among representations. For example, upper elementary school students should develop facility in moving back and forth among the concrete and the pictorial models, the oral name, and the symbolic representation of any fraction or decimal. These connections are vital if students are to make sense out of later operations on numbers.

Emphasizing representation at the secondary school level is a greatly needed instructional improvement. Whenever possible, a variety of representations of concepts should be presented, explored, or suggested. For example, the idea of linearity needs to be represented in tabular form so that students can observe change numerically; it needs to be represented graphically so that a visual and geometric representation is familiar; and it needs to be represented symbolically so that symbolic meaning as well as visual meaning can be attached to ideas such as slope and intercept (see fig. 1.2). A similar position is appropriate for other classes of functions as well. Solid and connected understanding is based on comfort with a variety of representations.

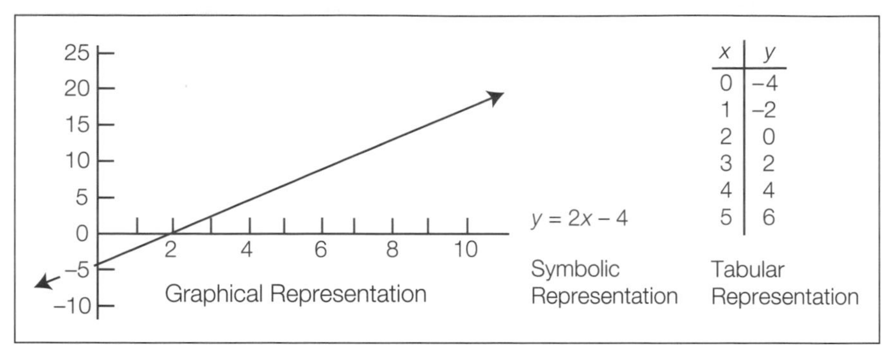

Fig. 1.2

In geometry and data analysis, multiple representations are also vital. Data need to be looked at, organized into tables and frequency charts, and represented graphically in a variety of ways to develop a connected view of data and data analysis. Similarly, in geometry the synthetic, picture-drawing approach needs to be connected and related to other representations using coordinates, vectors, and equations. For example, a reflection in a horizontal line can be introduced synthetically; a coordinate rule can be developed for reflection in the x-axis, $(x, y) \rightarrow (x, -y)$; and the matrix representation of the reflection can then be developed.

Applications, Problem Solving, and Reasoning

Applications, problem solving, and reasoning are discussed together, since there is little disagreement about their importance in developing a connected view of mathematics. Certainly problem solving and reasoning have been mainstays of mathematics instruction for a long time, and applications have recently been revived. Applications can help to connect mathematics and the learner. Areas of interest to large numbers of students can be the basis for data-analysis projects or for modeling situations with functions such as linear or exponential variations. For example, data analyses can be done on popular recording artists, cars, or fashions that apply the mathematics in a student-friendly context. Conversely, these same contexts can be used to capture the attention of students, thus motivating them to develop mathematical concepts, a methodological suggestion that should be considered seriously by teachers at all levels.

Applications and problem-solving activities need to provide opportunities for multiple mathematical approaches so that students do see the connections. As an example, consider the following situation (see fig. 1.3):

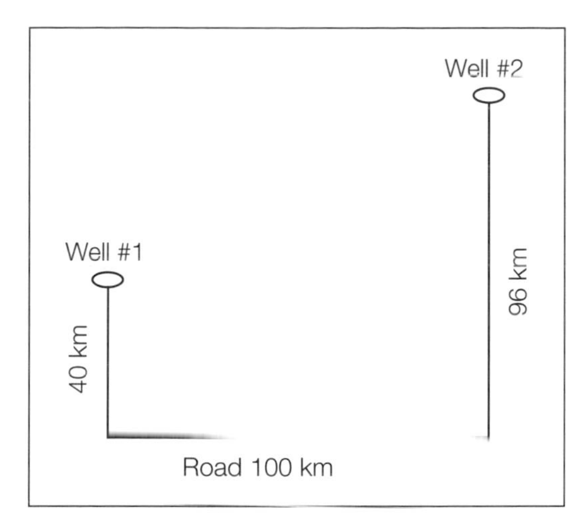

Fig. 1.3

> Two natural-gas wells are located 40 km and 96 km due north of points 100 km apart on a straight east-west highway. The gas company is planning to build a distribution center somewhere on the 100 km of highway. Since pipe and labor are very expensive, the gas company wants to build the center in a place that minimizes labor and material costs. Where do you suggest the center be built? Support your suggestions.

This problem lends itself to a number of solution paths. It could be done with scale drawings, with a geometry drawing program, by coordinatizing and graphing algebraic models, or by using geometric reasoning and the triangle inequality. By sharing their different solutions, students again emphasize the interrelatedness of the mathematical ideas and their representations. Note also that mathematical reasoning is required in the problem. These processes need to be regular features of all the mathematics students do to ensure that they develop the habit of providing mathematical evidence and understand the need for mathematical justification of the positions they take.

In a very real sense, the mathematical-processes aspect of mathematical connections forms a continuous web of emphasis. The four processes identified here should occur regularly in mathematical instruction. Students should see them used and use them in relation to all content strands, and teachers should point to their use regularly. In this way, the whole of mathematics is seen as interrelated and connected.

CONNECTORS

The final aspect of mathematical connections discussed here is that of mathematical connectors. Mathematical connectors include ideas such as *function, matrix, algorithm (procedure), graph, variable, ratio,* and *transformation.* They are mathematical ideas that arise in relation to the study of a wide spectrum of topics. As such, they permit the student to see the use of one idea in many different and, perhaps, seemingly unrelated situations.

In this yearbook, the idea of transformations as a connector is developed in chapters 7, 8, and 9. The notion of function as a connector is developed in chapter 6 and in the NCTM Addenda Series book *Connecting Mathematics* (Froelich 1991). Function is especially useful in connecting mathematics to the external world, since many real-life situations are modeled by functions.

Algorithms

The idea of an algorithm or procedure is one that is pervasive in much that we do. In the elementary school, much time is spent on having students investigate and develop ways to perform operations systematically. The results of these investigations are algorithms or procedures for computing, for telling time, for constructing place-value numerals, and for many other everyday aspects of mathematics. Such activity in the modern curriculum furnishes opportunities to exercise mathematical reasoning as well, since the algorithm should not be accepted without rationale and justification appropriate to the developmental level of the students.

The increase in the importance of computer technology has heightened interest in algorithms. Once a procedure is developed by students and validated mathematically, that procedure is ready to be programmed so that a calculator or computer can perform it. This helps the student see the connections of mathematics to the external world and also helps counteract fear of the mysterious workings of the computer.

Programs that draw and manipulate geometric shapes, such as the Geometer's Sketchpad (1990) or Geometry Inventor (1992), offer another way to see the power of algorithms. Many programs allow scripts to be written that will draw other figures using the procedures developed while constructing an initial figure. This is a powerful application of algorithms and the usefulness of technology. Thus, the idea of *algorithm* or *procedure* is a connector that can be used in much of mathematics.

Graphs

Graphs seem to be another kind of connector, especially with the world of real accomplishments and setbacks. At early levels, graphs can be used to represent many everyday data sets. Weekly graphs summarizing attendance, achievements, games won in sporting events, students bringing lunches, and so on provide an obvious connection of mathematics to life. Graphs thus illustrate a mathematical representation that may clearly display the important information in a situation.

Later in the curriculum, graphs are used to represent solutions to equations or inequalities; to represent functions and relations; to represent problem situations; to display data visually so that trends and tendencies can be observed; to represent patterns found in all strands; and, in discrete mathematics, to serve as an object of study and to model a variety of situations. Graphs supply a dimension to the mathematics not readily available in other representations, namely, the overall visual impression of the nature of a distribution, a function, or a pattern. It is a connector that needs greater emphasis in our instructional plans to counterbalance the heavy reliance on abstraction and symbolization in much of what is done. Graphs also help students with different learning styles connect to the mathematics more easily.

Variable

Variable is another pervasive connector. *Variable* has several meanings, from the unknown in a problem to the changeable argument in a function to the pure symbol found in statements such as the distributive property. In elementary school mathematics, seeking solutions to equations such as $3 + \triangle = 9$ or $5 - 2 = \triangle$ introduces students to one meaning of variable. Using an expression such as $8 \cdot x$ to generate a pattern 8, 16, 24, ... is another use of variable. Calculator technology gives rise to the idea of variable as a location storing a number. In data analysis, variables offer a way to connect the mathematics with the world of the learner through data collection and the analysis of situations interesting to the students. The variables may represent heights, hours of TV watched, number of toys owned, and so on. In geometry, variable assumes a subtle, yet powerful role in mathematical reasoning. In proof situations, when we "let $\triangle ABC$ be an isosceles triangle" or use some other such statement, the variable is $\triangle ABC$, and it represents the class of isosceles triangles. Thus, even though we view a sketch of one isosceles triangle, the proof is for the class of all such triangles in the same way that the proof of

$$ab = ac \Rightarrow b = c, \text{ with } a \neq 0 \text{ and } a, b, c \text{ being real numbers,}$$

is a proof for all real numbers. An examination of the content being taught is likely to reward you with more ways in which you can show the connectedness of mathematics using the *variable* connector.

Ratio

The final connector proposed here is ratio. It is useful at nearly all levels of mathematical study. In the elementary school, the idea of ratio is introduced and used in situations such as two cookies to a student at a party or the ratio of students with brown eyes to those with other colors. Since ratios often are generated out of situations familiar to students, ratio yields another way to connect the students to mathematics. Ratios are found in percents in the later grades of elementary school, and questions about ratios lead naturally to the idea of π as a ratio. Ratio is found in many problem-solving situations, such as in rate, time, and distance relations. Ratio is found in geometry in size transformations and in similarity as the ratio of similitude. In its common development, probability is defined as the ratio of successful trials to total trials. Ratio is obvious in the concept of slope and in the trigonometric functions. In nearly any content or problem situation, the notion of ratio is evident and can thus act as a connector for the mathematics.

FINAL WORDS

In this article we have attempted to illustrate, admittedly briefly, three aspects of mathematical connections that we called unifying themes, mathematical processes, and mathematical connectors. We tried to suggest that each category furnishes opportunities for students to experience the connectedness of mathematics if only they are given the chance and, perhaps, a friendly nudge to think more broadly than just in terms of the topic at hand. It is clear that students will need a "friendly nudge" quite often, and that that nudge often will need to come from a teacher who sees not only the trees but the forest as well. Thus, the teacher must be prepared to emphasize the connections through discussions in class, through probing questions asked during presentations of solutions, through asking students to look for relations and seek connections, and through the creation and selection of problems and learning activities that allow a variety of mathematical approaches. As students and teachers continue to "think connections," the connectedness of the mathematics will grow and become dominant. When that occurs, all will wonder why anyone had ever thought of mathematics in any other way.

BIBLIOGRAPHY

Commission on the Secondary School Curriculum of the Progressive Education Association, Committee on the Function of Mathematics in General Education. *Mathematics in General Education.* New York: D. Appleton-Century Co., 1940.

Froelich, Gary. *Connecting Mathematics.* Addenda Series. Reston, Va.: National Council of Teachers of Mathematics, 1991.

Geometer's Sketchpad. Key Curriculum Press, Berkeley, Calif., 1990.

Geometry Inventor. Wings for Learning, Scotts Valley, Calif., 1992.

National Committee on Mathematical Requirements of the MAA. *The Reorganization of Mathematics in Secondary Education.* Buffalo, N.Y.: Mathematical Association of America, 1923.

National Council of Teachers of Mathematics. *Curriculum and Evaluation Standards for School Mathematics.* Reston, Va.: The Council, 1989.

2

Connections as Problem-Solving Tools

Theodore R. Hodgson

RECENTLY, a friend of mine declared that the learning of mathematics is a contact sport. "In order to become proficient," she said, "students must make contact with the ideas of mathematics, search for the answers *and* the justifications, and assume an active role in their own learning processes." In many ways, these observations lie at the heart of current efforts to reform mathematics education. The learning of mathematics is an active and constructive process (Skemp 1987; Noddings 1990). Students interpret classroom activities in light of their existing beliefs and assimilate information into their existing knowledge structures. As a result, each student constructs a kind of "personalized" mathematics. To some, mathematics is a collection of isolated rules and facts. However, mathematics can also be perceived as a network of ideas in which each idea is connected to several others.

According to the *Curriculum and Evaluation Standards for School Mathematics* (National Council of Teachers of Mathematics [NCTM] 1989), students should seek to connect the ideas of mathematics. Specifically, the authors of the *Standards* contend that the establishment of connections among mathematical concepts enables students to appreciate the power and beauty of the subject. Furthermore, it is claimed that an ability to form and use connections empowers students as problem solvers (p. 146):

> Students who are able to apply and translate among different representations of the same problem situation or of the same mathematical concept will have at once a powerful, flexible set of tools for solving problems.

One goal of this yearbook is to identify activities that promote the use of connections in problem solving. However, prior to addressing instructional issues, one should first understand what connections are and what impact they have on the problem-solving process. In the NCTM *Standards,* mathematical connections are characterized as problem-solving "tools." In this article, the meaning of that metaphor is examined and evidence is offered that supports the NCTM's view of connections. Subsequently, some classroom implications of the "connections as tools" analogy are identified.

ON CONNECTIONS AND THEIR USES

Constructivists maintain that all knowledge is personal and arises from our active attempts to interpret the world around us (Noddings 1990). Say the word *tool* to ten different people, and in all likelihood they will conjure up ten different images. This article seeks to identify the connection between connections and tools. However, in order to ensure that all readers interpret *tool* in a similar manner, it is essential that we first establish a common working definition of the word. In this article, *tool* represents an implement or object used to perform an operation or to carry on work of any kind (*Webster's New Third International Dictionary*).

According to this definition, any object is a tool as long as it is used by somebody or something to accomplish some objective. Naturally, the first objects that come to mind when the word *tool* is mentioned are physical implements. Calculators and computers are tools of the mathematics classroom because they are used by students to explore and learn mathematics. However, objects need not be physical to be tools. For example, even though they cannot be touched, words are often considered tools because they are the objects by which speakers convey ideas. As long as an object is used in some way to achieve an end, it is a tool.

Like speech, connections among processes or concepts in mathematics are objects in the abstract sense of the word. As an illustration, consider the connection between the equation $y = x - 1$ and the graph in figure 2.1. If solutions to the equation are plotted on an xy-plane, the result is the graph in figure 2.1. Thus, the graph is actually a visual representation of solutions to the equation and, indirectly, of the equation. Conversely, the equation is a symbolic representation of the graph.

Obviously, the connections between graphs and equations are not physical objects. Rather, the connections exist in the minds of individuals. For instance, I have plotted the solutions to the preceding equation and compared the results to the graph in figure 2.1. In my mind, the graph and the equation are related. However, if an individual fails to identify correspondences between the graph and the equation, then, for that individual, each remains an isolated piece of mathematical information.

The fact that connections are constructs of the mind raises an interesting and widely debated question. Are concepts in mathematics (such as graphs and equations) inherently connected or do the connections exist only in the minds of learners? Although this is an interesting philosophical question, to me the answer is irrelevant. If students are unable to establish connections, then the connections cannot be used in problem situations regardless of whether they exist or not. However, once they are established in the mind of the learner, connections are "things" that can be retrieved from memory and, if the need arises, applied to problems. It is for this reason that connections are considered problem-solving tools; they are objects that are used by problem solvers to solve problems.

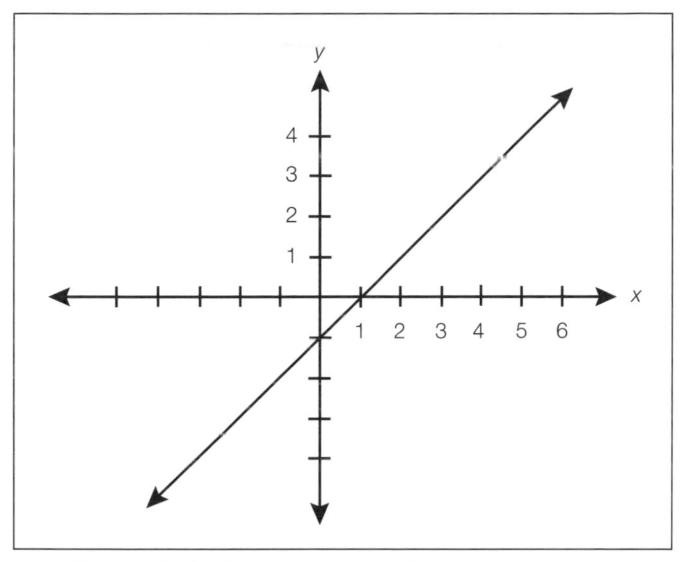

Fig. 2.1. A visual representation of the equation $y = x - 1$

To illustrate the role of connections as problem-solving tools, consider one approach to solving the following integral expression:

$$\int_{x=1}^{3} (x - 1)\, dx$$

If the equation $y = x - 1$ is graphed and if it is recognized that the value of the definite integral is equal to the area of the shaded region in figure 2.2, then the problem can be solved by applying an appropriate area formula. In this approach to the problem, the solution follows from the use of two symbolic-to-visual translations: (1) the graphing of $y = x - 1$, and (2) the identification of the region of the plane corresponding to the integral expression. Of course, it is possible to solve the problem directly, without the use of visual representations. However, the establishment of connections between symbolic and visual representations offers an alternative path to the solution.

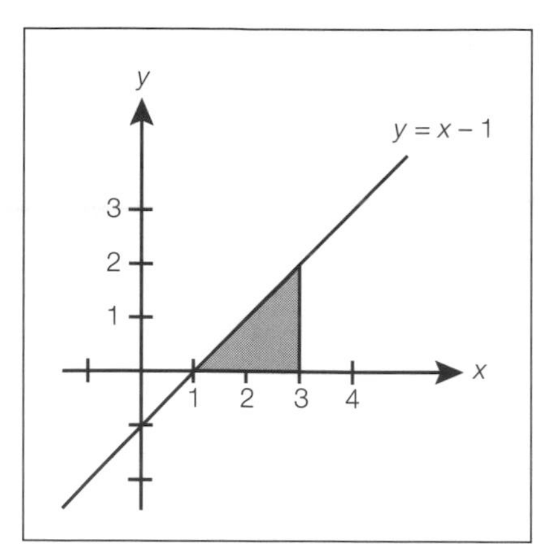

Fig. 2.2. A visual representation of the area corresponding to the designated integral

As an additional example, consider the traditional sequence of steps by which multibase rods and units are used to complete the following subtraction task:

$$\begin{array}{r} 32 \\ -\ 19 \\ \hline \end{array}$$

By design, the trade-in value of rods and units (one rod for ten units) mirrors the relationship of adjacent columns of base-ten numbers. However, the convenient design of multibase blocks does not by itself ensure that students will connect the blocks to base-ten numbers. To solve the problem using the blocks, students must first establish correspondences between the blocks and digits in the appropriate columns. Units must be used to represent digits in the ones column; rods must be used to represent digits in the tens column; and students must recognize that the relationship between rods and units parallels the relationship between digits in the tens and ones columns, respectively. On the basis of these connections, problem solvers construct a concrete representation of the minuend (see fig. 2.3a), trade one rod for ten units (fig. 2.3b), remove one rod and nine units (fig. 2.3c), and construct a symbolic representation of the remaining rods and units (fig. 2.3d). In addition, the blocks and numbers must be connected on an operational level. Students must identify an action (the removal of rods and units) that corresponds to the operation of subtraction.

In these two examples, connections enable students to construct alternative representations of problem situations and to operate within these alternative problem spaces to find solutions. For instance, the connections formed between multibase blocks and base-ten numbers allow students to construct a concrete representation of the initial problem, to operate in that concrete world in order to obtain a solution, and to translate back

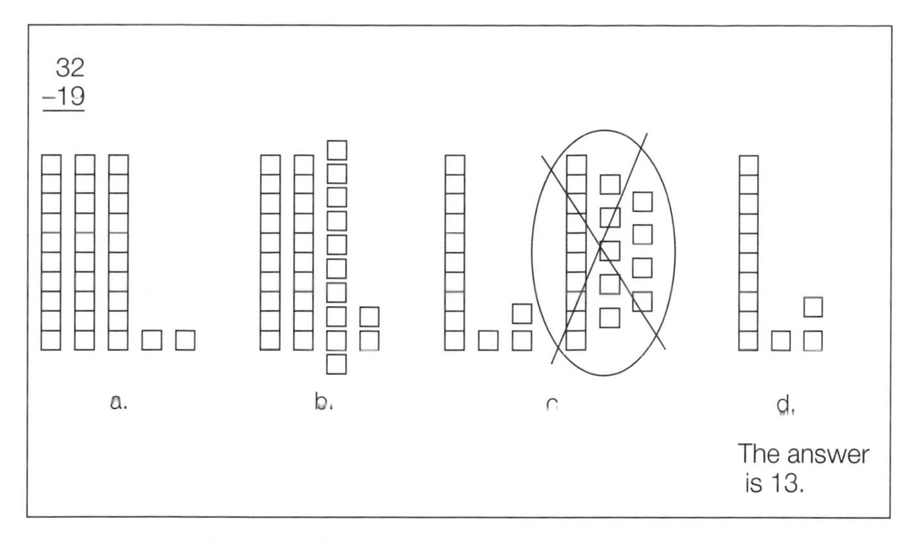

Fig. 2.3. Using multibase blocks to evaluate the expression 32 − 19

from the concrete world to the symbolic. Of course, this is not the only way in which connections are used. The recognition that two algorithms (such as the standard subtraction algorithm and the cashier's method of finding the difference by counting up from the subtrahend) achieve the same end allows problem solvers to substitute one algorithm for another but may not alter the initial representation of the problem. However, the examples do illustrate some uses of connections.

As further evidence of the connections-tools relationship, consider the essential role of problem solvers in the problem-solving process. In general, tools are incapable of completing tasks independently; they must be used by operators to do so. Likewise, connections cannot solve problems directly, but they enable problem solvers to do so. For example, although connections can lead to the completion of the integration task discussed earlier, problem solvers must still identify the goal of the problem, understand how connections relate to this goal, apply the connections in an appropriate manner, and recall and apply an appropriate area formula. In other words, connections merely facilitate the completion of problems. They are tools used by problem solvers to solve problems.

Some Implications for Instruction

Since connections are problem-solving tools, the teacher's task is to promote the use of connections in problem solving. In a sense, this is a twofold task. Classroom instruction must enable students (1) to establish connections, and (2) to apply the connections in problem settings. To understand better the distinction between these two skills, consider the following analogy to tennis. In tennis, a player's strokes—forehands, backhands, lobs, and so forth—are all tools for returning the ball over the net. Situations arise that call for a particular stroke. However, if the player has never practiced (and *constructed*) the stroke, then he or she will be unable to return the ball. This is analogous to the establishment of connections: in order to use connections to solve problems, students must have established the connections. They must consciously identify correspondences among mathematical ideas and connect the ideas in their minds.

In addition to perfecting the basic strokes, tennis players must also learn to apply each one in game situations. For instance, if my opponent is charging the net, then my best chance of winning the point may be to lob the ball over his or her head. If he or she is out of position at the net, however, then my best bet is to hit a forehand or backhand into the open court. The best tool to use depends on the situation, and it is the ability to recognize which tool is best that distinguishes experts from novices. The same is true of connections. Certain problems seem to lend themselves to particular connections; thus, students should learn to recognize problem situations that call for the establishment of connections and to apply the connections in those situations. In the remainder of this article some thoughts are offered regarding the means by which teachers can accomplish this dual task of preparing students to establish and use connections.

A Problem-Solving Approach

As is illustrated throughout this yearbook, mathematics abounds with concepts and processes that are or can be connected. As teachers, we must be aware of such relationships and draw on them to solve problems. However, the learning of mathematics is a "contact sport." We cannot simply transfer our perceptions and problem-solving expertise to students; we must engage them in activities that require them to construct ideas for themselves. For that reason, an understanding of connections should emerge from the exploration of problem situations.

As an example, students in many elementary school classrooms investigate real-world problems and collect, analyze, and interpret real-world data. What a perfect place to illustrate the uses of connections! To solve their problems, students must establish connections between the original problem and the collected data. Additionally, because raw data are generally difficult to interpret, students often establish visual representations of their data through the use of bar graphs, stem-and-leaf plots, or pie charts. Lastly, students must interpret their findings in light of the original problem.

In general, the investigation of problem situations leads naturally to the establishment and use of connections. In turn, the use of connections to solve problems brings about the need for their establishment. Connections are not seen as merely interesting mathematical facts but as integral components of successful problem solving.

The Importance of Follow-Up Activities

Ideally, the use of real-world problems actively involves all students and promotes the establishment and use of connections. However, simply engaging students in activities that use connections does not ensure that students will become proficient in their use. The discussions that follow in-class activities are equally important. Students should reflect on their own problem-solving efforts, as well as the efforts of other students, and identify the connections that were used. They should state the goal of the problem, identify the role connections played in achieving that goal, seek alternative solution strategies, and explore the use of other connections.

Follow-up activities should also address the techniques underlying the establishment of the connections. Students should identify correspondences between mathematical concepts or processes and understand how to translate from one representation to another.

Finally, students must be held accountable for their understanding of connections. Assessment activities should investigate students' abilities to establish and use connections, and feedback should allow students to correct any misunderstandings.

The Establishment of a "Connections Tool Kit"

Classroom instruction should also focus on the connections among a variety of representations. For instance, the recent emphasis on mathematical modeling requires students to sift through verbal descriptions of problem situations, identify relevant problem variables, collect and graph data regarding those variables, identify algebraic relationships from the visual displays, and interpret the results in terms of the original problem. Thus, modeling activities lead students to construct, among others, visual representations of numeric expressions, algebraic representations of visual displays, and contextual interpretations of algebraic or numeric expressions.

An ability to establish and use a wide range of connections offers students alternative paths to the solution. In particular, with the formation of each new connection, the tool kit that students bring to each problem is expanded and the likelihood of discovering a solution is enhanced. As an analogy, the worker who is proficient with more than one tool is more likely to complete the task at hand. If one tool doesn't work, he or she simply reaches into the tool kit and tries another.

Connections as a First Step

Since connections serve as tools in the problem-solving process, issues regarding the use of connections are, in fact, issues about problem solving. Research and experience have taught us one simple fact: problem solving is an extremely difficult skill to analyze and to teach. In his famous book on problem solving, Polya (1973) identified a four-step process that characterizes successful problem solving. First, successful problem solvers develop an understanding of the problem and construct a plan of attack. In many instances, the problem-solving plan calls for the establishment and use of connections. However, as the following scenario illustrates, problem-solving plans can also emerge from the establishment of connections.

The "handshake problem" is well known to mathematics educators. Simply stated, one must determine the number of handshakes that occur in a room containing a given number of people if each person shakes hands with every other person exactly once. In my initial attempt to solve the problem some years ago, I used twenty-five circles to represent twenty-five people and drew lines to represent handshakes. To obtain the answer, I would count the number of lines. However, as figure 2.4 illustrates, I soon discovered the difficulty of this "brute force" approach to the problem. As lines were added to the diagram, the complexity of the figure soon became unmanageable. As a result, I devised an alternative problem-solving plan: Construct similar but simpler problems and look for a numeric or algebraic pattern.

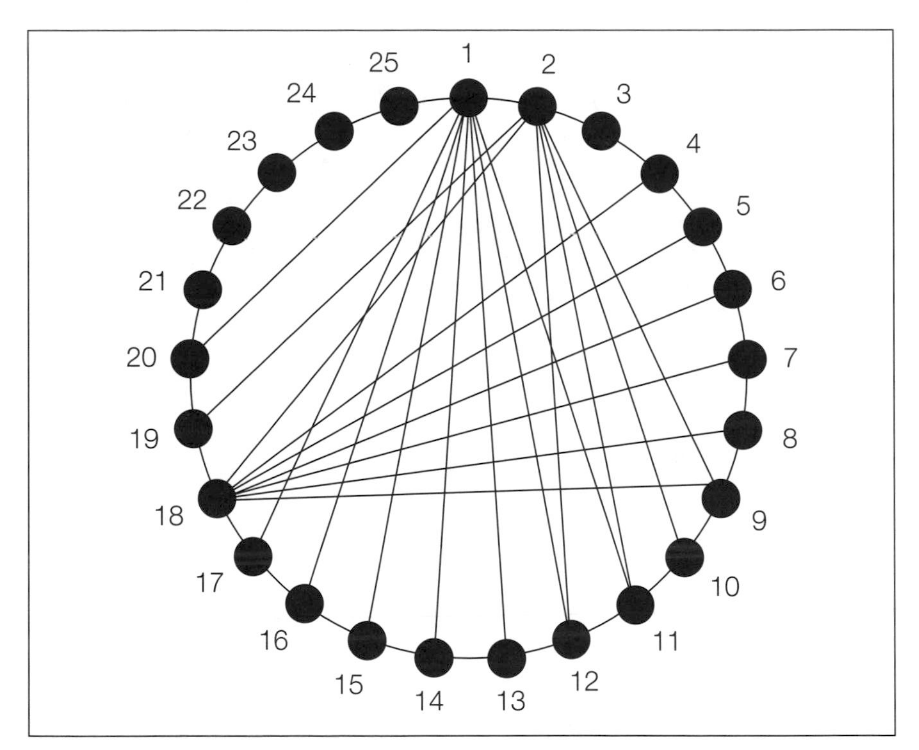

Fig. 2.4. A partial visual representation of the "handshake problem"

In reference to Polya's terminology, I understood the problem and developed a plan that involved the construction of a visual representation of the problem information (a connection between visual and natural-language representations). However, in implementing the plan, I discovered it to be too complex and time-consuming. Thus, the establishment and use of the connection led to an alteration of the initial plan.

In summary, an understanding of the problem and the development of a plan need not precede the use of connections in problem solving. In their book on the teaching of problem solving, Charles and Lester (1982) recommend that students establish connections, through the use of lists, tables, charts, pictures, objects, or graphs, to understand a problem better. In other words, the establishment of connections can lead to a better understanding of a problem and, ultimately, to successful problem solving.

CONCLUSIONS

The authors of the *Standards* claim that mathematical connections serve as tools in the problem-solving process. On the basis of the arguments and evidence presented in this article, it appears that this analogy is justified. An ability to establish connections enables students to solve problems. However, the establishment of connections does not guarantee problem-solving success. The impact of connections on the problem-solving process is user-dependent.

Assuming that connections serve as problem-solving tools, several instructional implications can be derived. First and foremost, this view of connections results in a dual instructional task: classroom activities must prepare students both to establish new connections and to use established connections in problem settings. If students are unable to establish connections, then connections can never play a role in problem solving. Likewise, an inability to use established connections renders their establishment irrelevant. Thus, the accomplishment of both of these objectives underlies successful problem solving.

REFERENCES

Charles, Randall, and Frank Lester. *Teaching Problem Solving: What, Why, and How.* Palo Alto, Calif.: Dale Seymour Publications, 1982.

National Council of Teachers of Mathematics. *Curriculum and Evaluation Standards for School Mathematics.* Reston, Va.: The Council, 1989.

Noddings, Nel. "Constructivism in Mathematics Education." In *Constructivist Views on the Teaching and Learning of Mathematics,* edited by Robert B. Davis, Carolyn A. Maher, and Nel Noddings. *Journal for Research in Mathematics Education* Monograph Series, no. 4. Reston, Va.: National Council of Teachers of Mathematics, 1990.

Polya, George. *How to Solve It.* 2nd ed. Princeton, N.J.: Princeton University Press, 1973.

Skemp, R. R. *The Psychology of Learning Mathematics.* Hillsdale, N.J.: Lawrence Erlbaum Associates, 1987.

3

Connecting School Science and Mathematics

Donna F. Berlin
Arthur L. White

THE integration of school science and mathematics has received much attention in current education reform documents as a means for improving student performance and understanding and for developing realistic and positive attitudes and perceptions related to science and mathematics. A plethora of terms can be found in the literature to refer to "integration," including *connections, cooperation, coordination, correlated, cross-disciplinary, fused, interactions, interdependent, interdisciplinary, interrelated, linked, multidisciplinary, transdisciplinary,* and *unified* (Berlin 1991). Throughout the literature, there is a general sense that integration is a "good" thing. However, very little has been reported that explicitly describes what it means to integrate science and mathematics, and even less research has been done to explore its benefits or detriments (Berlin 1991). Although many might agree with the proposition "integrate how you teach before worrying so much about integrating *what* you teach" (Steen 1994), others advocate the infusion of "mathematical methods into science and scientific methods into mathematics such that it becomes indistinguishable as to whether it is mathematics or science" (Berlin and White 1992, p. 341). All this points to the critical need to develop a common language through the elaboration of a model for the integration of school science and mathematics.

INTEGRATION MODEL

The integration of school science and mathematics needs to attend to a broad range of aspects to guide educational practice and research effectively.

The writing of this document was supported by the U.S. Department of Education, Office of Educational Research and Improvement under grant #R117Q00062 to the Ohio State University for the National Center for Science Teaching and Learning. Any opinions, findings, conclusions, or recommendations expressed in this document are those of the authors and do not necessarily reflect the view of the sponsoring agency.

The Berlin-White Integrated Science and Mathematics (BWISM) Model identifies the following six aspects: (1) learning, (2) ways of knowing, (3) process and thinking skills, (4) conceptual knowledge, (5) attitudes and perceptions, and (6) teaching. We do not intend to imply that these aspects are isolated or exclusive of one another. The identification and elaboration of these as separate aspects is meant to provide some clarification of the characteristics that are in constant interplay as we seek to define, implement, and evaluate integration. These aspects, in various combinations, can serve as a basis to generate operational definitions and comparable research.

Learning

Integration can be viewed from the perspective of the learner and how scientific and mathematical concepts, processes, skills, and attitudes are processed and organized in the cognitive structure of the learner. Both science and mathematics educators value a constructivist view of learning (Piaget 1970; Vygotsky 1978) and the need to construct meaningful knowledge (Ausubel 1963; Novak and Gowin 1993). On the basis of cognitive research, the following general propositions may serve as the infrastructure for the development of integrated school science and mathematics:

- Knowledge is built on previous knowledge.
- Knowledge is organized around big ideas, concepts, or themes.
- Knowledge involves the interrelationship of concepts and processes.
- Knowledge is situation- or context-specific.
- Knowledge is advanced through social discourse.
- Knowledge is socially constructed over time.

These constructivist principles are manifested in all six aspects of the BWISM Model.

Recent findings from neuropsychology involving parallel processing in the brain may add a new and intriguing dimension to integration viewed from the learning perspective. In addition to serial information-processing capacities, humans have the capacity for parallel or simultaneous processing of information and events resulting in the unique ability to make predictions. The observations and experiences of individuals and the models, both qualitative and quantitative, that describe these events can be more efficiently assimilated when experienced simultaneously (Anderson 1992). If children experience qualitative and quantitative models at appropriate levels of abstraction from an earlier age, the ability to connect these models and make predictions should be enhanced.

Ways of Knowing

Science seeks to advance knowledge through the observation and manipulation of phenomena in order to explore the nature of the environment and human existence in that environment. Science searches for consistent and verifiable patterns to build a knowledge base and explain the real world. This way of knowing may be characterized as one of *induction* whereby a pattern or generalization emerges or is revealed from the data.

In mathematics, the quest for knowledge more frequently involves modeling in search of patterns and relationships that are not bound by the observable world and its relationships. The phenomena may not be readily observed or even envisioned because of extreme conditions such as size (e.g., very small or very large), time (e.g., past or future), or degree of abstraction. Mathematics often involves modeling and the use of logical, symbolic systems to describe patterns and relationships. These symbols can be manipulated without the constraint of reality or the need for concrete representation.

Figure 3.1 illustrates how the processes of induction and deduction are cyclically interrelated. The observation of the environment, collection of data, and search for relationships (induction) can deliver a basis for quantitative description. Mathematics can be used to express and communicate the relationships in the data as models (graphic, symbolic, numeric, geometric, or functional). These models may be subjected to mathematical transformations that can then be used to make predictions (deduction). Often the mathematical models can predict real-world phenomena that have not as yet been observed or even contemplated. For example, a

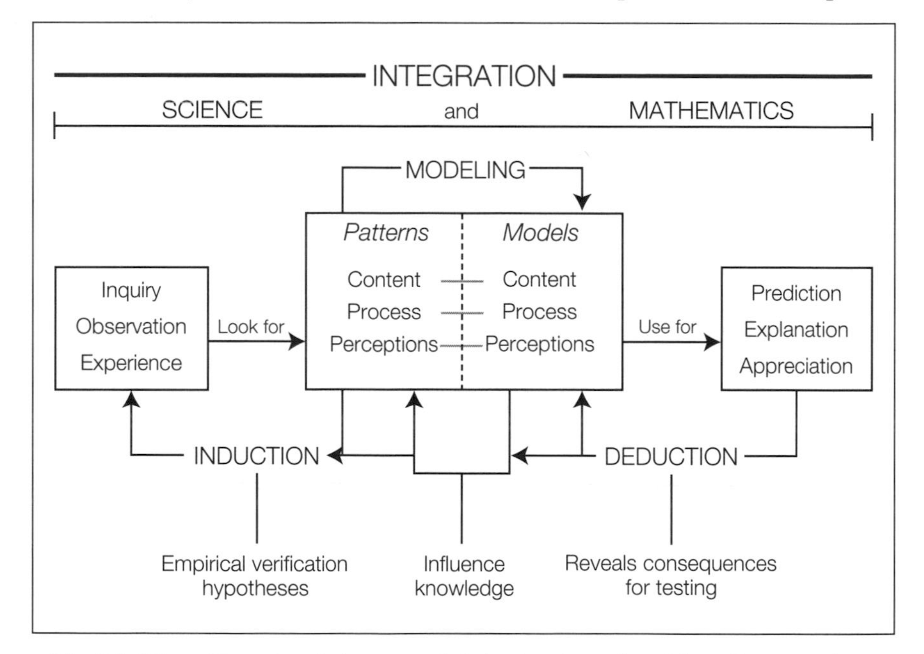

Fig. 3.1. Ways of knowing: induction and deduction (reproduced from Barnes et al. [1992], with permission from the National Center for Science Teaching and Learning, A. L. White, Director.

mathematical model could predict the existence of planets in the solar system or subatomic particles not yet discovered. These predictions can guide further exploration of the system within which the prediction was made, leading to additional observations and manipulations to verify or refute the predictions. If the predictions are verified, new knowledge and theories may result. If refuted, the results may be reconsidered mathematically and a revised model constructed.

The most significant feature of this "ways of knowing" cycle may be the importance of the symbiotic relationship between the inductive and deductive processes. It is a reasonable goal to provide ample opportunity for all students to acquire an understanding of the connections between these ways of knowing. If students are to benefit from the symbiotic relationship and power of these ways of knowing, the increased use of mathematical modeling in science classes and the use of student-generated scientific data in mathematics classes are recommended. Attention to this "ways of knowing" cycle aggregates the processes of induction and deduction to promote an integrated, holistic view of the generation of knowledge.

Process and Thinking Skills

Integration can be viewed from the perspective of process and thinking skills valued in both science and mathematics. Although the terms may differ, the mathematics curriculum standards recognizing problem solving (inquiry), reasoning, communication, and connections (integration) emerge as processes central to school science as well as mathematics (National Council of Teachers of Mathematics [NCTM] 1989; National Research Council 1993b, 1993c).

A review of the literature reveals widespread support for the inclusion of process (or inquiry) skills in science curricula. Science is a dynamic, ongoing, tentative search, and the process skills reflect the nature of science and the typical activity of scientists; they are readily transferred to other problem situations, and they generalize to real-life problems (American Association for the Advancement of Science 1963; National Research Council 1993b, 1993c; Padilla 1986). The National Science Teachers Association and numerous state and local curriculum specialists have emphasized process skills as specific, planned outcomes of science programs.

The development of basic and integrated science process skills has been a major emphasis of science education programs since the 1950s (Tobin and Capie 1980). The basic process skills include observing, inferring, measuring, communicating, classifying, and predicting. The integrated process skills include controlling variables, defining operationally, formulating hypotheses, interpreting data, experimenting, and formulating models. (See Padilla [1986] for definitions of these skills.) These same science process skills are integral to the development of problem-solving ability, which has been strongly endorsed as the primary focus of school mathematics (NCTM 1980, 1989).

Champagne (1992) has developed a comprehensive taxonomy of science and mathematics thinking skills that includes reasoning, metacognitive, information-manipulating, information-management, formal-reasoning, and symbolic-representation skills. Integrated science and mathematics activities have the potential to engage students in challenging, authentic, relevant problem-solving tasks that provide for the application and practice of higher-order thinking skills.

Conceptual Knowledge

Integration can be viewed from the perspective of the overlap of conceptual knowledge of science and mathematics. This requires an examination of the concepts, principles, and theories of science and mathematics to determine which ideas are unique to science or to mathematics and which ideas may overlap. An examination of current (NCTM 1989; Rutherford and Ahlgren 1990) and emerging (National Research Council 1993b, 1993c) documents on curriculum standards reveals a shared commitment to the development of a limited number of fundamental concepts. Topics common to both science and mathematics curricula include the study of measurement, patterns and relationships, probability and statistics, spatial relationships, and variables and function. Examples of ways to interrelate science and mathematics concepts could include (*a*) natural selection with sampling, randomization, probability, ratio, proportionality, and area; (*b*) activation energy and reaction rates with the area under the curve of a distribution; (*c*) population growth and genetics with sampling and probability; and (*d*) reflection and refraction with angular geometry.

Often concepts previously identified as "science" or "mathematics" can be related through analogous embodiments. For example, the variables that describe the properties of a lever and those that describe the characteristics of a frequency distribution are analogous and can link school science and mathematics. The balance point of a lever is the fulcrum. This is the point around which the clockwise and counterclockwise forces on the lever system are equal, or "balanced." Similarly, the mean of a distribution is the point in a frequency distribution around which the positive and negative deviation scores are equal and result in a "balance" point for the distribution. There are times when the concepts of science and the concepts of mathematics have enough in common for it to be more meaningful to the student and more efficient to connect the learning of these concepts (Penafiel and White 1989; White and Berlin 1989).

Attitudes and Perceptions

Integration can be viewed from the perspective of attitudes and perceptions related to the development of scientific and mathematical literacy. The beliefs and perceptions one holds about science and mathematics as disciplines may also be similar. Finally, creating an environment and providing opportunities to enhance one's scientific and mathematical self-efficacy may serve as another affective connection.

Current science- and mathematics-education reform documents are quite explicit about valued scientific and mathematical attitudes (NCTM 1989; National Research Council 1993b, 1993c). The following list (Loucks-Horsley et al. 1990, p. 41) has been modified to embody shared attitudes:

- Desiring knowledge: viewing science [and mathematics] as a way of knowing and understanding
- Being skeptical: recognizing the appropriate time and place to question authoritarian statements and "self-evident truths"
- Relying on data: explaining natural occurrences by collecting and ordering information, testing ideas, and respecting the facts that are revealed
- Accepting ambiguity: recognizing that data are rarely clear and compelling, and appreciating the new questions and problems that arise
- Being willing to modify explanations: seeing new possibilities in the data
- Cooperating in answering questions and solving problems: working together to pool ideas, explanations, and solutions
- Respecting reason: valuing patterns of thought that lead from data to conclusions and, eventually, to the construction of theories
- Being honest: viewing information objectively, without bias

Habits of mind or dispositions specific to current curricular, instructional, and assessment goals for both science and mathematics include curiosity, creativity, inventiveness, leadership, organization, persistence, resourcefulness, risk taking, self-confidence, self-direction, self-reflection, and thoroughness (NCTM 1989; National Research Council 1993b, 1993c; Rutherford and Ahlgren 1990; Stenmark 1991).

Another shared affective consideration is to promote science and mathematics as human enterprises. As such, they are approximations of truth and subject to change. Integrated experiences in science and mathematics may serve to dispel the notion of these subjects as unchangeable, irrefutable, proved bodies of knowledge that can furnish correct answers to all questions.

Integrated science and mathematics experiences based on personal and social issues and interests may motivate students to achieve. Opportunities to be successful help to encourage, support, and nurture students' confidence in their abilities to do science and mathematics. The promotion of scientific and mathematical efficacy may serve to counter the perceptions of science and mathematics as difficult, accessible to only a select group of individuals (e.g., Anglo males), and not essential to the general populace.

Teaching

Integration can be viewed from the perspective of science and mathematics teaching methods and strategies that overlap and are supportive of one another. Teaching, as related to integrated science and mathematics education, includes four dimensions: the structure and organization of the learning environment, instructional strategies, assessment, and the changing role of the teacher. The goal of integrated science and mathematics teaching is to enable students to "acquire both scientific [and mathematical] knowledge of the world and scientific [and mathematical] habits of mind at the same time" (Rutherford and Ahlgren 1990, p. 190). Such an environment would include a broad range of content, give time for inquiry-based learning, stimulate and support discourse, furnish opportunities to use laboratory instruments and other tools, provide appropriate and ongoing use of technology, encourage alternative assessment procedures, and maximize opportunities for successful experiences. Learning experiences should engage students in connecting qualitative and quantitative descriptions of their world. Modeling tools such as STELLA (Structural Thinking, Experimental Learning Laboratory with Animation, High-Performance Systems, Inc., Hanover, NH 03755) can help students to make the connections (Linn 1986).

Integrated teaching strategies should be based on cognitive research into how students learn in general and how they learn science and mathematics in particular. Initial activities should focus on interesting phenomena arising from, and applied to, real-world situations. There should be ample opportunity to develop thinking and reasoning skills; both basic and integrated process skills; and problem-solving, decision-making, and communication skills. Activities should involve students in doing, analyzing, and reflecting about science and mathematics and in acquiring scientific and mathematical power. A wide variety of school and community instructional resources involving multiple sensory modes (tactile, auditory, and visual) should be available. Cooperative-learning strategies are recommended to reinforce the collaborative nature of scientific and mathematical endeavors.

A variety of useful strategies to integrate across disciplines has been proposed by Fogarty (1991a, p. xv; 1991b, p. 63):

- Sequenced—topics or units of study are rearranged and sequenced to coincide with one another. Similar ideas are taught in concert while remaining separate subjects.
- Shared—shared planning and teaching take place in two disciplines in which overlapping concepts or ideas emerge as organizing elements.
- Webbed—a fertile theme is webbed to curriculum contents and disciplines; subjects use the theme to sift out appropriate concepts, topics, and ideas.
- Threaded—the metacurricular approach threads thinking skills, social skills, multiple intelligences, technology, and study skills through the various disciplines.

- Integrated—this interdisciplinary approach matches subjects for overlaps in topics and concepts with some team teaching in an authentic integrated model.

Alternative, authentic assessment procedures are suggested to meaningfully link integrated science and mathematics curriculum, instruction, and assessment. Embedded assessments may include performance assessments, projects, and portfolios. A number of current books related to alternative science (Hein 1990; Kulm and Malcolm 1991; Ostlund 1992; Raizen et al. 1989, 1990) and mathematics (Kulm 1991; Lesh and Lamon 1992; NCTM 1993; National Research Council 1993a; Romberg 1992; Stenmark 1989, 1991) assessments are available to assist classroom teachers in planning appropriate and meaningful experiences to assess what students know and to improve instruction.

Of significance to the teaching of integrated science and mathematics is the changing role of the teacher. Teachers and students become natural partners in problem solving and in developing their own capability for lifelong learning (NCTM 1989). The teacher may be viewed as an intellectual coach assuming such various roles as role model, consultant, moderator, interlocutor, and questioner (National Research Council 1991). Teachers as specialists and the use of teaching teams emerge in the integrated approach.

An Example: "Just Drop It!"

> In this investigation, students will compare the heights of bounces that result from dropping different types of balls from different heights. The bounces resulting from dropping the balls on different surfaces can also be compared. From the collected data, a formula relating the height of the bounce to the height of the drop will be developed. Using the formula, students will be able to accurately predict how high a ball will bounce from any given height.

This activity from *Math + Science: A Solution* (Wiebe and Ecklund 1987) illustrates the interdependence of the aspects of the integrated science and mathematics model. Children are involved in observing, experimenting, measuring, recording data, searching for patterns, interpreting, applying, predicting, and generalizing. The concepts and principles of science and mathematics developed in the activity include averages, coefficients, conservation, constants, elasticity, equations, friction, graphs, mathematical modeling, ratio, and slope.

The investigation can encourage and support instructional methods valued in both science and mathematics education. It is suggested that children work in cooperative groups of two to four while engaged in inquiry and problem solving. This methodology can promote the development of valued scientific and mathematical attitudes as well as positive habits of mind or dispositions. Other valued instructional methods embodied in this activity include hands-on manipulation of the environment; varied representational modes (e.g, drawings, diagrams, graphs, formulas); building on past experiences; the personalization of learning; the use of familiar and available resources (e.g., balls used in tennis, golf, basketball, baseball, soccer, pool, etc.); varied levels of questioning (ranging from those requiring low-level thinking to those requiring higher-order cognitive processing); the use of a simple, four-function calculator to average the data; and the use of a graphing calculator for data fitting and modeling. Consistent with the model, this activity furnishes numerous possibilities for making appropriate and effective connections between science and mathematics teaching and learning.

RESOURCES

Since the 1970s, a number of notable curriculum programs or supplements have been produced. Among these are the Minnesota Mathematics and Science Teaching Project (MINNEMAST—Minnemast Center, 720 Washington Avenue, S.E., Minneapolis, MN 55414; MINNEMAST materials are no longer in print, but they can be obtained from the ERIC Document Reproduction Service, CBIS Federal, 7420 Fullerton Road, Suite 100, Springfield, VA 22153-2852), Unified Science and Mathematics for Elementary Schools (USMES—Educational Development Center, Newton, MA 02160; USMES materials are no longer in print, but they can be obtained from the ERIC Document Reproduction Service, CBIS Federal, 7420 Fullerton Road, Suite 100, Springfield, VA 22153-2852), Activities Integrating Math and Science (AIMS—AIMS Educational Foundation, P.O. Box 8120, Fresno, CA 93747-8120), Great Explorations in Math and Science (GEMS—Lawrence Hall of Science, University of California at Berkeley, Berkeley, CA 94720), the Jasper Series (Optical Data Corporation, 30 Technology Drive, Warren, NJ 07059), Teaching Integrated Mathematics and Science (TIMS—University of Illinois at Chicago, Box 4348, Chicago, IL 60680), The Voyage of the Mimi and The Second Voyage of the Mimi (Sunburst/Wings, 101 Castleton Street, P.O. Box 100, Pleasantville, NY 10570-0100), and School Science and Mathematics Integrated Lessons (SSMILES—School Science and Mathematics Association, Bloomsburg University, 400 East Second Avenue, Bloomsburg, PA 17815-1301). Two documents to be published by the National Center for Science Teaching and Learning will soon be available as a resource for classroom teachers interested in integrated science and mathematics concepts, skills, and processes. The documents include a database describing approximately three hundred integrated science and mathematics activities (Berlin 1994) and a compilation of SSMILES (School Science and Mathematics Integrated Lessons) previously published in the journal *School Science and Mathematics* (Berlin in press). Materials related to consumer education, energy education, environmental education, nutrition education, technical education, and vocational education are also fertile areas for integrated science and mathematics activities.

CONCLUSION

If the implementation of integrated school science and mathematics is to occur in a rational and realistic fashion, it will be crucial to have teachers involved in the exploration and amalgamation of the six aspects as well as in gathering information on the effects of the integration. The model outlined in this paper can serve to guide educators, administrators, policy makers, and researchers in the development, implementation, and evaluation of school science and mathematics integrated activities and programs.

REFERENCES

American Association for the Advancement of Science. *Science—a Process Approach.* Washington, D.C.: The Association, 1963.

Anderson, O. Roger. "Some Interrelationships between Constructivist Models of Learning and Current Neurobiological Theory, with Implications for Science Education." *Journal of Research in Science Teaching* 29 (December 1992): 1037–58.

Ausubel, David P. *The Psychology of Meaningful Learning.* New York: Grune & Stratton, 1963.

Barnes, Marianne, Francis Conway, Lynn Narasimhan, Richard Shumway, and Arthur White. "Ways of Knowing: Induction-Deduction." In *Proceedings of the Integration Conceptualization and Writers Conference,* edited by Donna F. Berlin. Columbus, Ohio: National Center for Science Teaching and Learning, 1992.

Berlin, Donna F. *Integrating Science and Mathematics in Teaching and Learning: A Bibliography.* School Science and Mathematics Association Topics for Teachers Series, No. 6. Columbus, Ohio: ERIC Clearinghouse for Science, Mathematics, and Environmental Education, 1991.

———, ed. *Database of Integrated Science and Mathematics Instructional Activities.* Columbus, Ohio: National Center for Science Teaching and Learning, 1994.

———. *School Science and Mathematics Integrated Lessons (SSMILES).* School Science and Mathematics Association Classroom Activities Monograph Series. Columbus, Ohio: National Center for Science Teaching and Learning, in press.

Berlin, Donna F., and Arthur L. White. "Report from the NSF/SSMA Wingspread Conference: A Network for Integrated Science and Mathematics Teaching and Learning." *School Science and Mathematics* 92 (October 1992): 340–42.

Champagne, Audrey B. "Cognitive Research on Thinking in Academic Science and Mathematics: Implications for Practice and Policy." In *Enhancing Thinking Skills in the Sciences and Mathematics,* edited by Diane F. Halpern, pp. 117–33. Hillsdale, N.J.: Lawrence Erlbaum Associates, 1992.

Fogarty, Robin. *The Mindful School: How to Integrate the Curricula.* Palatine, Ill.: Skylight Publishing, 1991a.

———. "Ten Ways to Integrate Curriculum." *Educational Leadership* 49 (October 1991b): 61–65.

Hein, George, ed. *The Assessment of Hands-on Elementary Science Programs.* Grand Forks, N. Dak.: University of North Dakota Press, 1990.

Kulm, Gerald, ed. *Assessing Higher Order Thinking in Mathematics.* Washington, D.C.: American Association for the Advancement of Science, 1991.

Kulm, Gerald, and Shirley M. Malcolm, eds. *Science Assessment in the Service of Reform.* Washington, D.C.: American Association for the Advancement of Science, 1991.

Lesh, Richard, and Susan J. Lamon, eds. *Assessment of Authentic Performance in School Mathematics.* Washington, D.C.: American Association for the Advancement of Science, 1992.

Linn, Marcia C. *Education and the Challenge of Technology: Proceedings of a Conference on Technology and Teacher Education.* Cupertino, Calif.: Apple Computer, 1986.

Loucks-Horsley, Susan, Roxanne Kapitan, Maura D. Carlson, Paul J. Kuerbis, Richard C. Clark, G. Marge Melle, Thomas P. Sachse, and Emma Walton. *Elementary School Science for the '90s.* Andover, Mass.: NETWORK, 1990.

National Council of Teachers of Mathematics. *An Agenda for Action: Recommendations for School Mathematics for the 1980s.* Reston, Va.: The Council, 1980.

———. *Assessment in the Mathematics Classroom.* 1993 Yearbook of the National Council of Teachers of Mathematics, edited by Norman L. Webb. Reston, Va.: The Council, 1993.

———. *Curriculum and Evaluation Standards for School Mathematics.* Reston, Va.: The Council, 1989.

National Research Council. *Counting on You: Actions Supporting Mathematics Teaching Standards.* Washington, D.C.: National Academy Press, 1991.

———. *Measuring Up: Prototypes for Mathematics Assessment.* Washington, D.C.: National Academy Press, 1993a.

———. *National Science Education Standards: An Enhanced Sampler.* Washington, D.C.: The Council, 1993b.

———. *National Science Education Standards: July '93 Progress Report.* Washington, D.C.: The Council, 1993c.

Novak, Joseph D., and D. Bob Gowin. *Learning How to Learn.* Cambridge: Cambridge University Press, 1993.

Ostlund, Karen L. *Science Process Skills: Assessing Hands-on Student Performance.* Menlo Park, Calif.: Addison-Wesley Publishing Co., 1992.

Padilla, Michael J. *The Science Process Skills: Research Matters ... to the Science Teacher.* Calgary, Alta.: National Association for Research in Science Teaching, 1986. (ERIC Document Reproduction Service No. ED 266 961.)

Penafiel, Alfinio F., and Arthur L. White. "SSMILES: Exploration of the Mean as a Balance Point." *School Science and Mathematics* 89 (March 1989): 251–58.

Piaget, Jean. "Piaget's Theory." In *Carmichael's Manual of Child Psychology,* edited by Paul H. Mussen. New York: John Wiley & Sons, 1970.

Raizen, Senta A., Joan B. Baron, Audrey B. Champagne, Edward Haertel, Ina V. S. Mullis, and Jeannie Oakes. *Assessment in Elementary School Science Education.* Andover, Mass.: NETWORK, 1989.

———. *Assessment in Science Education: The Middle Years.* Andover, Mass.: NETWORK, 1990.

Romberg, Thomas A., ed. *Mathematics Assessment and Evaluation.* Albany, N.Y.: State University of New York Press, 1992.

Rutherford, F. James, and Andrew Ahlgren. *Science for All Americans.* New York: Oxford University Press, 1990.

Steen, Lynn Arthur. "Integrating School Science and Mathematics: Fad or Folly?" In *NSF/SSMA Wingspread Conference: A Network for Integrated Science and Mathematics Teaching and Learning—Conference Plenary Papers,* edited by Donna F. Berlin, pp. 7–12. Columbus, Ohio: National Center for Science Teaching and Learning, 1994.

Stenmark, Jean K., ed. *Assessment Alternatives in Mathematics: An Overview of Assessment Techniques That Promote Learning.* Berkeley, Calif.: EQUALS, 1989.

———. *Mathematics Assessment: Myths, Models, Good Questions, and Practical Suggestions.* Reston, Va.: National Council of Teachers of Mathematics, 1991.

Tobin, Kenneth G., and William Capie. "Teaching Process Skills in the Middle School." *School Science and Mathematics* 80 (November 1980): 590–600.

Vygotsky, Lev S. "Mind in Society." In *The Development of Higher Psychological Process,* edited by Michael Cole, Vera John-Steiner, Sylvia Scribner, and Ellen Souberman. Cambridge, Mass.: Harvard University Press, 1978.

White, Arthur L., and Donna F. Berlin. "SSMILES—Fulcrum and Mean: Concepts of Balance." *School Science and Mathematics* 89 (April 1989): 335–42.

Wiebe, Arthur, and Larry Ecklund. *Math + Science: A Solution.* Fresno, Calif.: AIMS Education Foundation, 1987.

4

Using Ethnomathematics to Find Multicultural Mathematical Connections

Lawrence Shirley

ONE of the logical extensions of the Connections Standard (National Council of Teachers of Mathematics [NCTM] 1989) is to encourage finding examples of mathematics in other cultures. The problem for many teachers is that most of the mathematics in our academic curriculum has been derived from the developments in European mathematics, and they have difficulty finding examples that do not seem Eurocentric.

Since our schools have increasingly heterogeneous populations from many different cultures around the world, an apparently European-based curriculum can be counterproductive to our interest in recruiting members of underrepresented groups into mathematics. If children see mathematics only from a European perspective, they may believe that non-European cultures have not worked with mathematics and, even worse, that these people *cannot* work with mathematics. If the child comes from a non-European cultural heritage, this belief may be dangerously extended to "*I* cannot work with mathematics."

Hence, we must actively seek instances of mathematics in other societies and mathematical developments from other cultures. This does not mean that we ought simply to find Chinese mathematics for our Asian students or Navajo examples for Native Americans or an Afrocentric curriculum for African American children. Rather, *all* students should be exposed to examples of mathematics from *all* cultures to see the applications and contributions of all and, more important, to respect the mathematical thinking of all.

One obstacle to finding such examples is that we tend to think of mathematical applications in the way we have always seen them. Since the mathematical applications of our own experience and the school mathematics we have seen in our own education often tend to be based on academic European mathematics, we need to look to other sources for examples. These may include history and geography books, anthropological sources, oral histories, the arts, or the applications of mathematics in the modern societies of other areas.

To assist us in looking for such uses of mathematics, we should broaden our view of what constitutes "mathematics." The diagram in figure 4.1 may help us to see areas of mathematics that deserve our consideration.

	Pure	Applied
Formal	academic	technical
Informal	recreational	everyday

Fig. 4.1. A broader classification of mathematics

It has long been common practice to divide mathematics into "pure" and "applied" categories. Figure 4.1 does that, but it also makes a significant cross-categorization, dividing mathematics into "formal" and "informal" mathematics. The result is a two-by-two grid of four cells, each of which will be described as we look at the implications for mathematical-cultural connections.

ACADEMIC MATHEMATICS

The upper left cell might be called *academic mathematics;* it is pure and formal and includes the fundamental laws, concepts, and techniques normally taught in schools and universities and continued in mathematical research. Much of the image—and the stereotype—of mathematics is derived from this cell, as are most of the examples, exercises, and problems of traditional mathematics curricula. Academic mathematics is also the focus of most histories of mathematics and, as described above, is often seen as *European* mathematics.

Some examples of non-European academic mathematics can be found in classical Chinese mathematics (Swetz and Kao 1977; Joseph 1991). Elementary school children learning about place-value numeration can benefit from seeing the systems of Chinese numeration that incorporated the place-value label into the numeral; for instance, 3486 would appear as 3 1000 4 100 8 10 6 in Chinese symbols (fig. 4.2). Children might compare this with the "Th-H-T-U" labels (for thousands, hundreds, tens, and units) that they sometimes use to help keep columns straight.

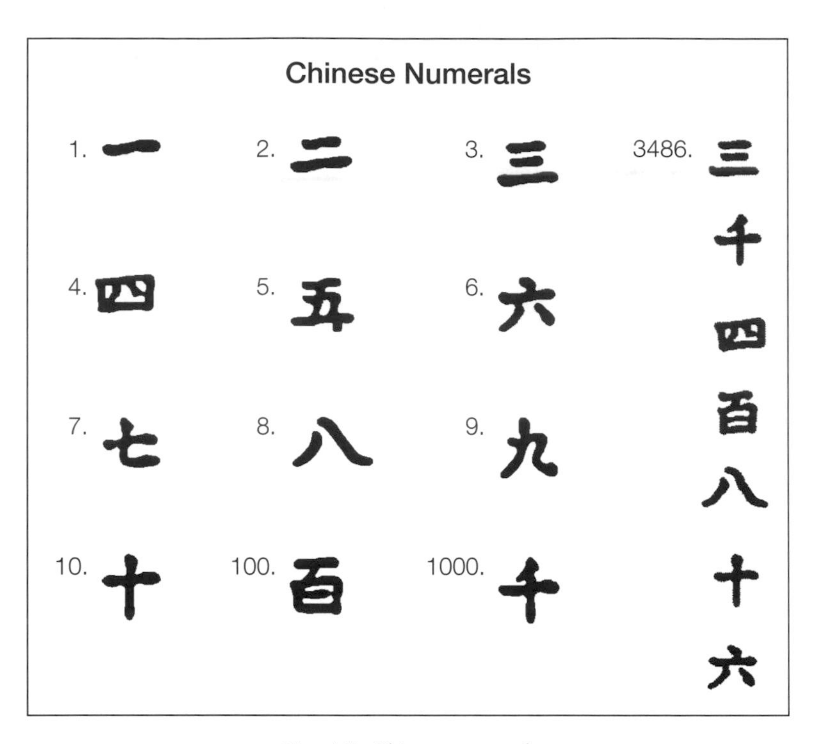

Fig. 4.2. Chinese numerals

High school students may be surprised to learn that the Chinese independently developed the "Pythagorean" theorem, the use of "Pascal's" triangle for binomial coefficients, and a technique for solving simultaneous equations similar to using matrices.

These examples represent a start toward broadening our view of mathematics. However, the academic cell is not the totality of mathematics. To look closely at mathematics, especially the mathematics of other cultures, we must consider the whole picture. The mathematics of other cultures may have come from many different directions, from many different sources, from many different styles of mathematical thought. To Western thinking, it may not even look like mathematics. The outward manifestation of technical skills, everyday chores, or recreational activities may cover complex mathematical relationships or sophisticated mathematical expertise. For example, Eglash and Broadwell (1989) reported fractal patterns in the layout of traditional Dogon villages in Mali, and Ascher (1991) noted a group structure in kinship relationships of the Walpiri people of Australia.

Ethnomathematics helps us to look at all four of the classification cells. D'Ambrosio (1985) has defined ethnomathematics as the mathematics of culture groups. The groupings may be based on ethnicity, but they may also be based on occupation (as in the mathematics of engineers, who do not worry about the rigors of proofs as much as about getting the job done correctly and efficiently), age (e.g., the mathematics in the video games of "kid culture," [Shirley 1991]), gender (Is there a difference in the mathematical thinking of women and men?), or others. It is important to look at all activities of the cultures—however they may be classified—to find mathematical patterns and structures. These may be so obvious that they have been ignored, or they may be deeply hidden in the nature of the activities. In any event, they can form a rich source of mathematical connections for classroom use.

TECHNICAL MATHEMATICS

The upper right cell in figure 4.1 is *technical mathematics*—formal and applied—including the upper levels of practical high school mathematics and some of college mathematics, as well as physics. Beyond the scope of our school mathematics, technical mathematics also includes mathematics of non-Western cultures used in carrying out some of the more complex tasks of their societies.

Students learning about compass directions in elementary school, angle bearings in middle school, or vectors and trigonometry in high school can appreciate, on several levels, the technical mathematical marvels of traditional Polynesian navigation on the open Pacific (Ascher 1991). Sailors of the Caroline Islands of the North Pacific take compass bearings from the rising and setting points of certain stars. Then they use a mental map of the islands' spatial relationships as reference points and an internal mathematical model that incorporates the winds and currents into an image of the islands moving past the canoe. From this modeling, they can guide their canoes across hundreds of miles of open ocean (fig. 4.3).

Another example of technical mathematics is the problem of distributing fish to inland village markets in West Africa (Gladwin and Gladwin 1971). The questions of supply and demand, transportation, and spoilage must all be considered. The people who fish and the people who sell must negotiate agreements on quantities, prices, profits, and the timing of deliveries. These discussions require mathematical thinking, strategies of problem solving, and a notion of geometric-network theory involving the locations of the markets and fishing areas, in addition to basic calculations.

Generally, technical mathematics is the mathematics of practical applications but is complex enough to require some kind of formal instruction. This instruction may not be given in schools, especially in non-Western cultures, but in an on-the-job, master-apprentice form of transmission of the concepts and, especially, the techniques.

EVERYDAY MATHEMATICS

Continuing around the grid to the lower right cell, informal and applied, we find *everyday mathematics.* As its name indicates, this type of mathematics takes its place in our everyday lives, in basic counting and arithmetic as well as in intuitive geometry, record keeping, and simple problem solving algebra. Children usually learn everyday mathematics in elementary school and refine it in middle school. However, in many traditional cultures without formal schooling, people require this mathematics for counting herds, handling market transactions, making farming decisions, and finding their way along networks of paths in the forest.

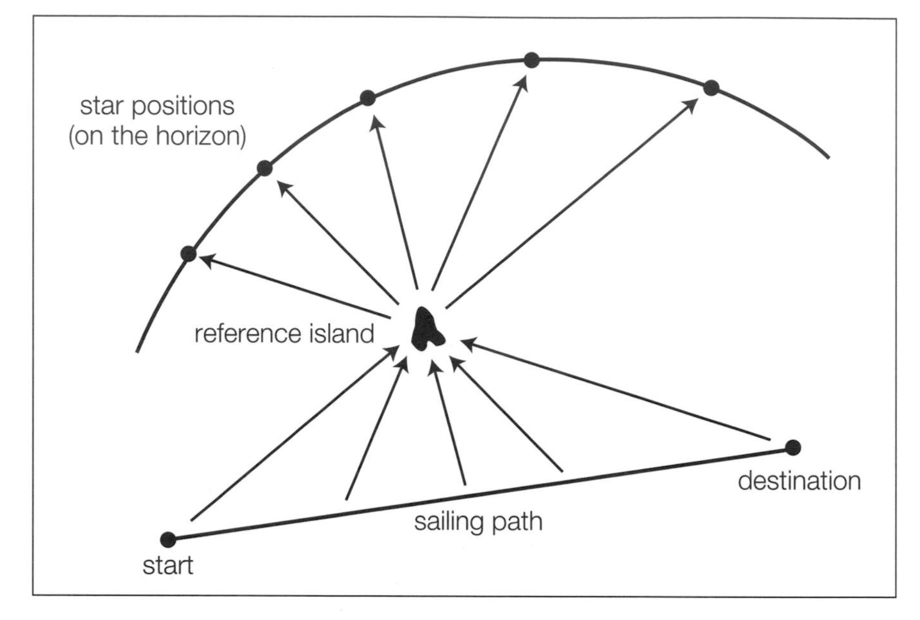

Fig. 4.3. Polynesian navigation (modified from Ascher [1991]). The "compass" of the Caroline navigators is the circle of the horizon, marked with the rising or setting points of thirty-two stars. Sailing is done by lining up a reference island with successive stars' compass positions as the journey proceeds. Of course, the sailors also must account for currents and wind, often sailing out of sight of land.

A surprising amount of this mathematical knowledge can be self-taught, and perhaps it is learned better from experience and immediate need than our school systems would like to acknowledge. A noted example (Saxe 1988) is that of Brazilian street children, most of whom have had little or no formal schooling. In selling candies and other items on the street, they smoothly and easily handle the sophisticated business arithmetic of profit and loss, bulk wholesale versus retail pricing, commissions, and even the widely fluctuating rates of inflation of the Brazilian currency, despite the lack of formal mathematical training.

In Nigeria, illiterate, unschooled adults, who need to use double-digit arithmetic for their daily transactions of farming and transportation, routinely take care of what we call the "tens column" before working with the smaller details of the units (Shirley 1988). It is interesting that this approach, opposite to that of our traditional algorithms, is currently being advocated for better understanding by second graders and for smoother operation of mental and written arithmetic (Kamii 1986).

Alan Bishop (1988) noted that every society, past and present, has had to find ways to carry out six kinds of activities: counting, measuring, locating, designing, explaining, and playing. All these activities lead to mathematical thinking and problem solving. Most belong in the cell of everyday mathematics, with "playing" probably reaching over to the recreational cell and the more complex levels of the others crossing up into technical mathematics. Learning about the vocabulary and structure of counting words or the systems of measurement units from around the world can enrich lessons on number and measurement. Geometry lessons can include methods of identifying locations and designing everyday tools. Explaining natural and societal processes can demonstrate mathematics in science and social studies lessons. By seeking examples of how different groups carry out daily activities, we can find mathematics in other cultures and demonstrate it in our classes.

RECREATIONAL MATHEMATICS

Recreational mathematics appears in the lower left corner of the classification grid. This mathematics is informal, often coming from games and puzzles, and it is pure, usually without a direct practical application beyond enjoyment and intellectual beauty. This is also the mathematics of art: symmetries, proportion, the golden mean, tessellations, and other patterns and visual relationships. Recreational mathematics includes rhythm, tones, choreography, and other mathematics of music. Sometimes this mathematics, which may appear irrelevant to "real" mathematics, can in fact lead to new mathematical research; for example, the harmonics of Pythagoras's music eventually led to Fourier analysis and the calculus of variations, early work in group theory made use of symmetry groups, and even esoteric ideas of graph theory found representation in puzzles such as the Königsberg bridge problem or the sand drawings of the Tchokwe people in Angola (Gerdes 1988).

This cell is rich in opportunities for classroom activities. One that is appropriate for all levels of school is the game generically called *mancala,* (also known as *oware, ayo, kalah, omweso, tei,* and many other names). Although the rules vary, all use a board of small "pits" containing "seeds," which can be any kind of small counters (see fig. 4.4). The play involves picking up the seeds from one pit and distributing them, one by one, sequentially into pits around the board. Certain special resulting situations, such as when the last seed of one's distribution lands on the opponent's side with a total of three seeds in the pit, allow seeds to be captured and removed from play. The player with the most seeds at the end of the game is the winner. The mathematics of the game requires simple skills of counting and remembering numbers as well as higher-level strategic thinking and problem solving (Haggerty 1964; Zaslavsky 1979; Shirley 1993).

Mancala

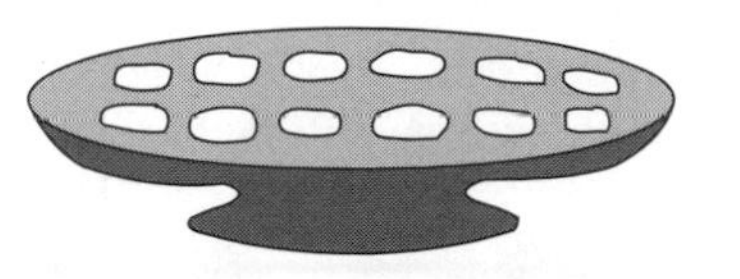

- *The board (four seeds in each pit):* An egg carton can be used, with buttons for seeds.
- *A play:* Pick up all the seeds in a pit on your own side and distribute them, one by one, into the pits (your own and your opponent's), counterclockwise around the board.
- *To capture seeds:* The last seed of your distribution must (*a*) land in a pit on your opponent's side, and (*b*) it must cause the total number of seeds in that pit to be either two or three; these seeds are removed and set aside.
- *The end of the game:* When one player has no more seeds, the other player keeps the seeds remaining on his or her side.
- *The winner:* The person with the most seeds when all have been captured wins.
- *The mathematics:* Players must keep track of the number of seeds in each pit (you can't pick them up to count them!) and the number needed to make captures; players also use strategic-thinking and problem-solving skills.

Fig. 4.4. The game of mancala

Examples of art from around the world can easily fit into class discussions of geometry, measurement, symmetry, and patterns. Native American and West African textiles show many types of symmetry, as do strip designs from South America and New Zealand. Classical Islamic art traditionally used regular and semiregular tessellation patterns, some of which became inspirations for the mathematical art of M. C. Escher. A sampling of art examples from around the world is shown in figure 4.5.

Examples of ethnomathematics may not easily be found without good access to the details of the cultures. Hence more research and, especially, a wider sharing of research information is necessary, even if the research is not initially intended as a search for ethnomathematics. Teachers and curriculum developers need to apply the theme of connections seriously to search for ideas. We can look at social studies texts, films, art, music, research, and popular articles that deal with other cultures. We must view these materials with a special eye, looking for mathematics where it may not be obvious. The task may be difficult, but a start has been made by a growing number of ethnomathematics research studies and books of examples based on these studies. Several are listed at the end of this article as a guide for teachers (see the entries in the References marked with an asterisk). More direct to classroom implementation is the increasing number of textbook examples and curriculum materials, including those that are not specifically intended for mathematics classes, based on multicultural themes.

As we approach the twenty-first century, the demands of the technological world require more people for mathematics: top-level scientists,

Adinkra Textile Patterns (Ghana)—symmetry

Islamic Tiles (Spain)—tessellations

Maori Rafter Designs (New Zealand)—symmetry

Navajo Rug (U.S.A.)—symmetry

Fig. 4.5. Mathematics in art around the world

engineers, and mathematicians who will continue the ongoing developments. The same demands call for more mathematics for the people: every citizen needs increased mathematical literacy to function in the modern world. To help satisfy these twin demands, the *Curriculum and Evaluation Standards for School Mathematics* (NCTM 1989) asks us to reach out with mathematics for all. To do this, to reach *all* students, we must consider the mathematics *of* all. Ethnomathematics, with its broader view of what mathematics is, can help us meet these needs.

BIBLIOGRAPHY

*Ascher, Marcia. *Ethnomathematics: A Multicultural View of Mathematical Ideas.* Pacific Grove, Calif.: Brooks/Cole Publishing Co., 1991.

Bishop, Alan. "Mathematics Education in Its Cultural Context." *Educational Studies in Mathematics* 19 (1988): 179–91.

D'Ambrosio, Ubiratan. "Ethnomathematics and Its Place in the History and Pedagogy of Mathematics." *For the Learning of Mathematics* 5 (1985): 44–48.

Eglash, Ron, and Peter Broadwell. "Fractal Geometry in Traditional African Architecture." *Dynamics Newsletter* 3 (4) (1989): 3–9.

*Gerdes, Paulus. "On Possible Uses of Traditional Angolan Sand Drawings in the Mathematics Classroom." *Educational Studies in Mathematics* 19 (1988): 3–22.

Gladwin, Hugh, and Christina Gladwin. "Estimating Market Conditions and Profit Expectations of Fish Sellers at Cape Coast, Ghana." In *Studies in Economic Anthropology,* Anthropology Studies No. 7, edited by George Dalton. Washington, D.C.: American Anthropological Association, 1971.

*Haggerty, John B. "Kalah—an Ancient Game of Mathematical Skill." *Arithmetic Teacher* 11 (May 1964): 326–30.

*Joseph, George Gheverghese. *The Crest of the Peacock: Non-European Roots of Mathematics.* London: Penguin Books, 1991.

Kamii, Constance. *Young Children Continue to Reinvent Arithmetic—Second Grade.* New York: Teachers College Press, 1989.

*Krause, Marina C. *Multicultural Mathematics Materials.* Reston, Va.: National Council of Teachers of Mathematics, 1983.

National Council of Teachers of Mathematics. *Curriculum and Evaluation Standards for School Mathematics.* Reston, Va.: The Council, 1989.

*Nelson, David, George Gheverghese Joseph, and Julian Williams. *Multicultural Mathematics: Teaching Mathematics from a Global Perspective.* New York: Oxford University Press, 1993.

Saxe, Geoffrey B. "Candy Selling and Math Learning." *Educational Researcher* 17 (August-September 1988): 14–21.

Shirley, Lawrence. "Historical and Ethnomathematical Algorithms for Classroom Use." Paper presented at the Sixth International Congress of Mathematical Education, Budapest, Hungary, 1988.

*———. "Mathematics: African Style." *Australian Mathematics Teacher* 49 (3) (1993): 24–25.

*———. "Video Games for Math: A Case for 'Kid Culture.'" *ISGEm Newsletter* [International Study Group on Ethnomathematics] 6 (2) (1991): 2–3.

*Swetz, Frank J., and T. I. Kao. *Was Pythagoras Chinese? An Examination of Right-Triangle Theory in Ancient China.* Reston, Va.: National Council of Teachers of Mathematics, 1977.

*Zaslavsky, Claudia. *Africa Counts: Number and Pattern in African Culture.* Brooklyn, N.Y.: Lawrence Hill Books, 1979.

*———. "People Who Live in Round Houses." *Arithmetic Teacher* 37 (September 1989): 18–21.

*———. "Symmetry in American Folk Art." *Arithmetic Teacher* 38 (September 1990): 6–12.

PART 2

Connections within Mathematics

5

Connecting Number and Geometry

Lowell Leake

CONNECTIONS—what connection can be more important than that between number and geometry? This article hopes to show how teachers in the early grades can build that connection as they teach the addition and multiplication facts. In the process, other connections will appear between number and algebra and between two-dimensional and three-dimensional space. As teachers in the earliest grades make the connections suggested here, they can also make connections with teachers in later grades, including high school, who can follow up and expand the ideas for students at all levels. The connections between spatial and numeric ideas become increasingly sophisticated as children work through the mathematics curriculum. This article illustrates such growth in sophistication, beginning with the addition and multiplication facts.

The basic idea is quite simple—three-dimensional addition and multiplication tables. Typically, the traditional addition and multiplication tables are introduced as in figure 5.1, and students memorize the basic number facts. In addition to teaching the facts using traditional memorization methods, what if we also physically built three-dimensional models of the tables using a column of, for example, six blocks to represent either the *4* + *2* cell in the addition table or the *3* × *2* cell in the multiplication table?

Let's start with the addition table. Put a large copy of the empty addition table on the desk or floor and have an ample supply of some kind of blocks available. The blocks in the accompanying photographs are Unifix cubes; Multilink blocks would work as well. (The commercially available manipulatives that snap together are probably the easiest blocks for a classroom teacher to use.) The idea is for children to put the appropriate number of blocks on each cell in the table instead of writing the symbol

---

The author thanks two of his colleagues for their assistance: Timothy J. Hodges for supplying the Mathematica drawing of a saddle surface and Robert M. Drake for providing the manipulatives used in some of the photographs.

+	0	1	2	3	4	5
0	0	1	2	3	4	5
1	1	2	3	4	5	6
2	2	3	4	5	6	7
3	3	4	5	6	7	8
4	4	5	6	7	8	9
5	5	6	7	8	9	10

×	0	1	2	3	4	5
0	0	0	0	0	0	0
1	0	1	2	3	4	5
2	0	2	4	6	8	10
3	0	3	6	9	12	15
4	0	4	8	12	16	20
5	0	5	10	15	20	25

Fig. 5.1. Addition and multiplication tables

for the number in the cell. Figure 5.2 shows all the columns of blocks for the sums up to 5 + 5. A straw has been placed across the tops of columns that are five units high to emphasize that pairs of numbers that add up to five lie in a straight line; the same would be true for any fixed sum. Piles as high as ten units become precarious; as a practical matter, it may be better to avoid manipulating sums greater than seven or eight and instead produce pictures that show students what it would look like if they use higher sums. Figure 5.3 shows a three-dimensional spreadsheet graph that serves this purpose. (All the spreadsheets in this article were produced by Excel version 4.0 [1992].) Although the use of spreadsheets may not be appropriate until middle school or later, their use offers another opportunity to draw connections within the mathematics curriculum.

As students mature mathematically, they discover that there are numbers between the whole numbers—rationals and, later, irrationals. There is no

Fig. 5.2

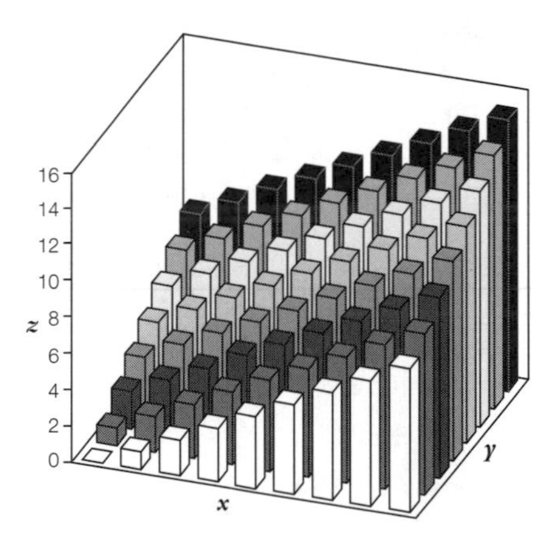

Fig. 5.3. Addition table: whole numbers

room in the traditional tables for such problems as 1/2 + 3/4. To circumvent this obstacle, we need to think of addition and multiplication tables in a different way. Figure 5.4 shows the alteration needed: when numbers other than integers are involved, lattice points, where two lines cross, must be used, not cells.

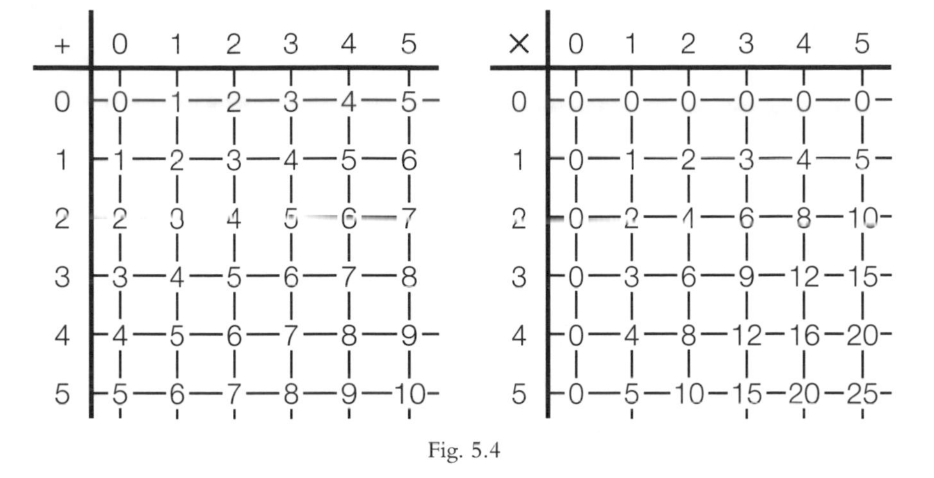

+	0	1	2	3	4	5
0	0	1	2	3	4	5
1	1	2	3	4	5	6
2	2	3	4	5	6	7
3	3	4	5	6	7	8
4	4	5	6	7	8	9
5	5	6	7	8	9	10

×	0	1	2	3	4	5
0	0	0	0	0	0	0
1	0	1	2	3	4	5
2	0	2	4	6	8	10
3	0	3	6	9	12	15
4	0	4	8	12	16	20
5	0	5	10	15	20	25

Fig. 5.4

Perhaps this is the way addition and multiplication tables should be first introduced to emphasize that other numbers exist between two whole numbers and that numbers do not jump from 2 to 3, for example, without passing through something else. If the three-dimensional representation is expanded to include all nonnegative rationals (and irrationals), the tops of the columns form a smooth surface. Figure 5.5 shows this surface; it forms part of a plane where the height of the plane at any point equals the sum of the two numbers beneath.

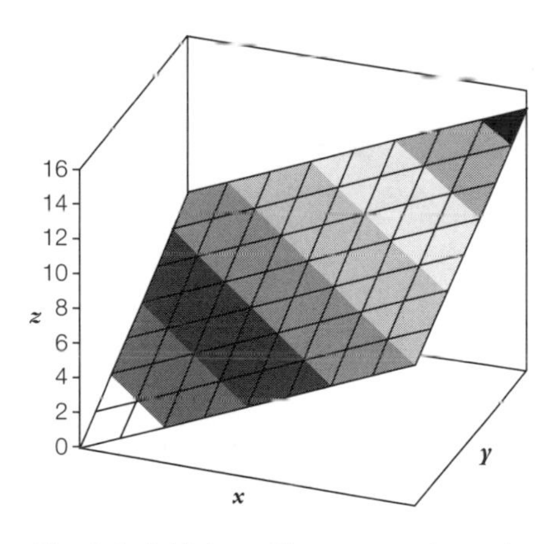

Fig. 5.5. Addition table: nonnegative reals

When students first encounter negative numbers, the difficulty of representing negative sums as piles of blocks on a table or on the floor emerges. How can a teacher or students pile blocks below the floor or under the desk to illustrate, for example, the equation −5 + 3 = −2? The columns for negative sums must extend in the opposite direction. Figure 5.6 shows the nonnegative sums for a limited number of integers; figure 5.7 shows how the model can be extended to an addition table for all the integers from −5 to 5.

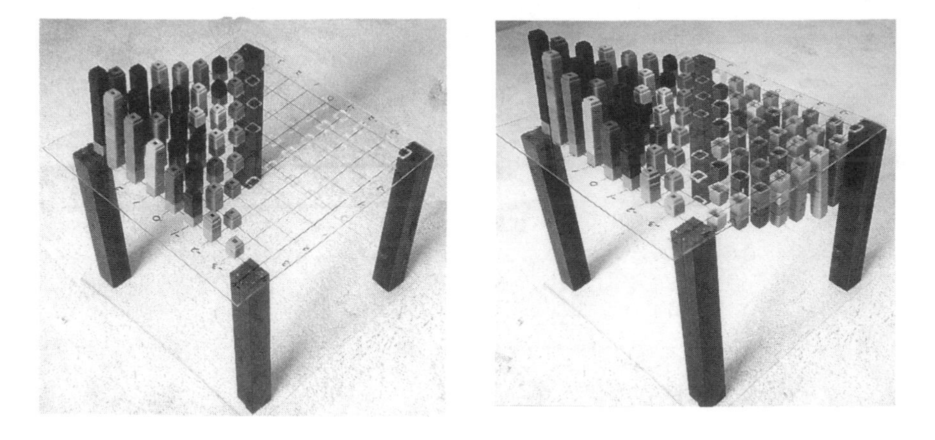

Fig. 5.6

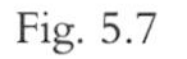

Fig. 5.7

In figure 5.7 the blocks have been glued to the plastic sheet. Note that the strings of blocks that extend upward (positive sums) are separated from those that extend downward (negative sums) by a line of sums that all equal 0. In an ordinary graph in the xy-plane, this would be the line $y = -x$ because when one number is the negative of the other, the sum of the two is 0. Figure 5.8 shows the three-dimensional spreadsheet picture of the addition table for the integers. Once again, such spreadsheets add a further sophistication to the main idea presented in this article. Figure 5.9 shows an extension of this idea for all real numbers, using a spreadsheet.

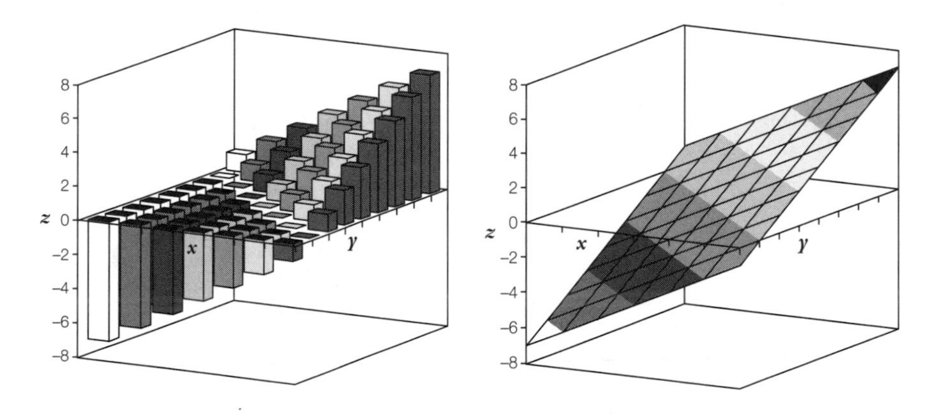

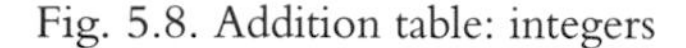

Fig. 5.8. Addition table: integers

Fig. 5.9. Addition table: reals

The addition table appears ultimately as a plane that slants its way through space. In fact, the addition table is simply the graph of $f(x,y) = x + y$. Students who work their way through grades K–12 with this kind of early experience, reinforced along the way, and who understand that the value of

$f(x,y)$ is simply the height of the surface above or below the xy-plane should have less trouble in high school and college classes understanding graphs of surfaces in three-dimensional analytic geometry or in multivariate calculus.

A simple example of $f(x,y) = x + y$ is to think of buying x pieces of candy at one cent each and y pieces of bubble gum at one cent each. The equation $f(3,4) = 3 + 4 = 7$ shows the total cost (7 cents) of 3 pieces of candy and 4 pieces of bubble gum. An example that is more sophisticated is, "How much would you spend on x shirts that cost \$10 each and y pairs of slacks that cost \$19 dollars each?" In this example, $f(x,y) = 10x + 19y$. This equation also represents a plane in space; however, it tilts differently from $f(x,y) = x + y$. One can even make sense of negative integer values for x or y if one thinks of returning merchandise for credit. For example, what would the bill be if you bought five shirts but returned seven pairs of slacks? An even more sophisticated application of the idea of $f(x,y) = x + y$ is the following: "If x is the number of \$3 tickets to a concert and y is the number of \$5 tickets and if you pay \$2 to park your car, the total amount you spend will be $f(x,y) = 3x + 5y + 2$ (assuming only one car is used)." This equation also represents a plane, which also tilts differently from $f(x,y) = x + y$ and is raised by two units at the origin. Once again, if you allow the return of tickets for credit, it makes sense to have positive or negative integer values for x and y.

What about the multiplication table? Figure 5.10 shows the three-dimensional multiplication table for a series of whole numbers using Unifix cubes. The tops of the blocks form a curved surface, not a plane. Note that piles of blocks of the same height are arranged in curves (called hyperbolas) of the form $x \cdot y = k$, where k is the value of a particular product. In figure 5.10 this is illustrated by the bent wire along the tops of piles that are six units high, formed from the different factors of 6. Figure 5.11a shows the spreadsheet version of the same table, and figure 5.11b shows the smoothed surface for the nonnegative real numbers.

Extending the multiplication table to include fractions and the negative numbers produces the same difficulties encountered in the addition table. Lattice points must be used instead of cells, and negative multiples must be represented by strings of cubes that extend below the xy-plane. Figure 5.12 shows the result using the same technique as for the addition table. Strings of blocks above the plastic sheet represent positive products, and strings below it represent negative products. Notice that strings of blocks above the plastic sheet are separated from strings of blocks below the plastic sheet by *two* lines of products equal to zero. Figure 5.13 shows the spreadsheet version.

Figure 5.14 shows the spreadsheet surface created if all real numbers are used, and figure 5.15 shows the same surface drawn with Mathematica (1992). The surface for the multiplication table is one of the most fascinating in all of mathematics—a hyperbolic paraboloid, the famous saddle surface—first encountered in calculus. The equation of the hyperbolic paraboloid pictured in figure 5.15 is simply $f(x,y) = x \cdot y$. This is

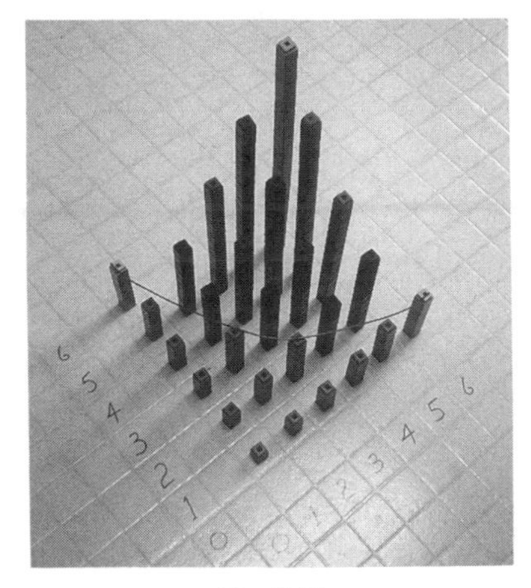

Fig. 5.10

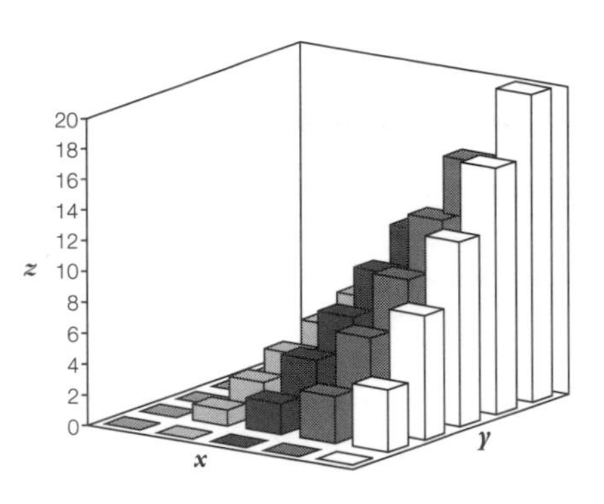

Fig. 5.11a. Multiplication table: whole numbers

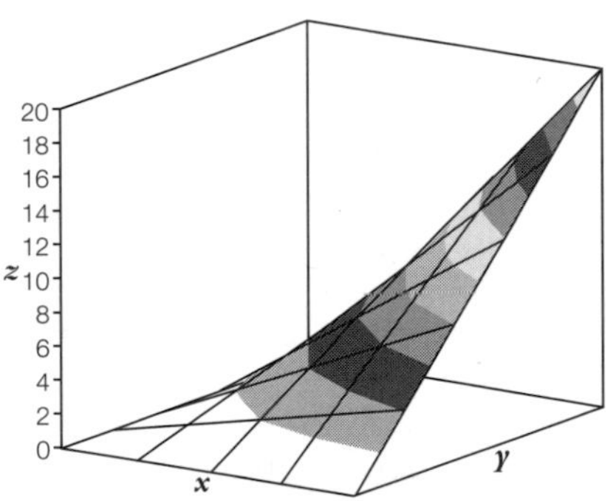

Fig. 5.11b. Multiplication table: nonnegative reals

Fig. 5.12

the surface usually introduced in calculus as $f(x,y) = x^2 - y^2$, but that becomes $f(x,y) = x \cdot y$ if it is rotated 45 degrees around the vertical axis. It is interesting to notice that when children learn their multiplication facts, they are dealing with the underlying idea of a hyperbolic paraboloid. So the multiplication table is a saddle; it rises up fore and aft (for the products of two positives or two negatives) and it rolls downward to the sides (for the products of a positive and a negative).

The hyperbolic paraboloid is a doubly ruled surface; at any point on the surface there exist two straight lines intersecting at that point that lie entirely on the surface. This makes for some interesting constructions. Figures 5.16a and 5.16b show such a homemade surface. To construct such a surface, make a wooden rectangular frame (the one shown is square) hinged on opposite corners and drill equally spaced holes along the

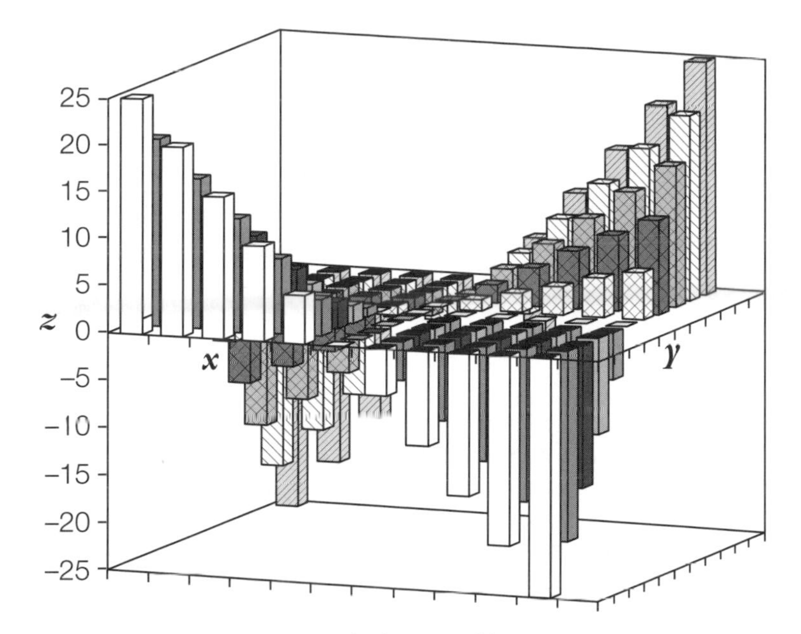

Fig. 5.13. Multiplication table: integers

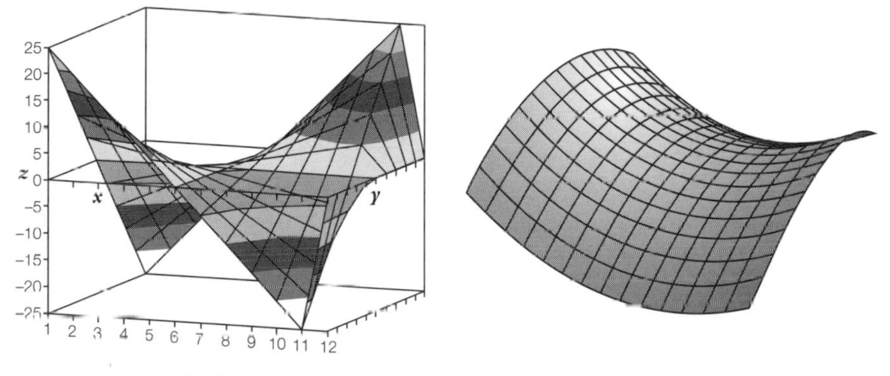

Fig. 5.14. Multiplication table: reals

Fig. 5.15

edges. Lift two opposite corners so that the pairs of adjacent sides form a dihedral angle of about 90 degrees (fig. 5.16). This can be a permanent arrangement or a hinged one as in figure 5.16. Then thread a cord of good quality between the holes on opposite edges until taut. If the frame is hinged, it must be held rigid while you thread the cord. (This was done with a plastic rod in figure 5.16a.) The strings are the lines that lie entirely on the surface, and there are two such strings for each point on the surface. The hinges allow you to collapse the model and put it in a suitcase or a drawer for traveling or storage.

Fig. 5.16a

Fig. 5.16b

If you look around your town, you may find such a hyperbolic paraboloid used as the roof of a building. The surface is very popular with builders and architects for roofs of buildings because, to the architect, it is a very appealing and artistic curved surface and, to the builder, it can be constructed with straight beams. Figure 5.17 is a photograph of such a building, Bellarmine Chapel at Xavier University in Cincinnati, Ohio.

As I was gluing the Unifix cubes to the plastic sheets, two questions came to mind that might challenge high school students and their teachers. One question arose from the large number of cubes that were needed to build the model of the addition and multiplication tables: "How many cubes would be needed to build an addition table for the positive whole numbers that range from 0 + 0 to, for example, 15 + 15, or to $n + n$?" This is the same as asking, "How much do the sums in the addition table, from 0 + 0 to 15 + 15 (or $n + n$), add up to?" The question can be answered by brute force for a specific n, or it can be done by summing certain series. Doing the same thing for the multiplication table is similar but a bit more difficult.

Fig. 5.17

The second question arose when I was converting the addition-table model to the multiplication model. Since a lot more gluing was involved, I wondered, "Are there any combinations of x and y whose sums are the same as their products?" This is the same as asking, "For what values of x and y does $x + y = x \cdot y$?" The geometric interpretation of this question is, "Where does the surface $f(x,y) = x + y$ intersect the surface $f(x,y) = x \cdot y$?" Perhaps some teachers and students in high school will investigate these questions.

Readers should know that there is a better way than gluing blocks to the plastic sheets for negative sums or products. I have developed a model for doing this with Unifix cubes. Simply draw a lattice diagram on a transparent plastic sheet about one foot square with lattice lines about 1.5 inches apart. Then glue a flat square piece of plastic that fits into the open end of a Unifix cube on each lattice point on each side of the plastic sheet. This means you will need about 130 of the flat squares. This will produce a plastic sheet with square "warts" on each side, at the lattice points, to which Unifix cubes can be attached. The flat squares of plastic have to be measured and cut professionally. For Multilink cubes they need to be a different size and shape. Gluing them to the sheet takes time and patience. Figure 5.18 is a photograph of my model. Perhaps the companies who make cubes can be persuaded to include such a piece of equipment in the packages they sell to schools or teachers.

Fig. 5.18

To summarize, the simplest addition and multiplication facts of elementary school arithmetic can lead to sophisticated connections among an extraordinary variety of mathematical concepts and their applications, and the computer can play an important role. Although the computer cannot replace the actual concrete experiences of using strings of Unifix cubes or Multilinks blocks to represent three-dimensional addition and multiplication tables, it *can* build on those concrete experiences, and what the computer ultimately produces is made much more meaningful when students have had concrete experiences with cubes and blocks.

REFERENCES

Excel. Version 4.0. Redmond, Wash.: Microsoft Corp., 1992.

Mathematica. Version 2.1. Champaign, Ill.: Wolfram Research, 1992.

6

Using Functions to Make Mathematical Connections

Roger P. Day

IN THE *Curriculum and Evaluation Standards for School Mathematics,* the National Council of Teachers of Mathematics (NCTM) observed that "one of the central themes of mathematics is the study of patterns and functions" (NCTM 1989, p. 98). Standards for grades K–4, 5–8, and 9–12 reflect that observation, emphasizing students' explorations of patterns and relationships. The *Standards* suggests establishing a strong foundation for the function concept using informal investigations in the primary and middle grades with extensions to formal symbolism and the discussion of functions in high school.

Functional relationships offer fertile ground for making mathematical connections. As a unifying idea in mathematics, the function concept helps students connect different mathematical ideas and procedures. Functional relationships also provide connections to other content areas and a perspective from which to view real-world phenomena. To illustrate, this article presents connections derived from classroom explorations of functional relationships. To review the historical development of functions, see Kleiner (1989), and for results of research into how students learn about functions, see Markovits et al. (1988) and Vinner and Dreyfus (1989).

CLASSROOM EXAMPLES

Primary Grades

Many topics of interest to students stem from functional relationships. A primary-grade teacher may ask students to tell the number of pets in their families. The unique number of pets that corresponds to a family illustrates a function. Asking students to discuss the impossibility of one family having two different numbers of pets reinforces the uniquely determined correspondence. In addition, asking students to generate their own examples of functional relationships (such as each student's unique height, age in months, or eye color) helps students realize how frequently such relationships occur in their lives.

A functional relationship also furnishes a context within which primary-grade students can make mathematical connections. A data set for part of a class, generated through the activity described above, is shown in table 6.1. Questions of number sense and operations spring from the functional relationship represented in the data set:

- Whose family has fewer pets, Lisa's or Ashton's?
- How many more pets are in Usha's family than in Kirk's?
- How many pets will Denny's family have if his brother brings home four rabbits?
- Name the students whose families have the same number of pets.
- How many pets are there among all the families?

TABLE 6.1
Number of Pets in Students' Families

Name	Number of Pets	Name	Number of Pets
Reinhold	3	Gloria	0
Lisa	4	Elke	1
Lu	2	Lyuba	3
Michiko	1	Denny	6
Sam	0	Alejandro	2
Lolita	0	Kirk	1
George	2	Margie	3
Usha	3	Ashton	2
Claire	8	Torri	1
Donovan	2	Yed	0

Other questions that depend on the data provide a connection to probability, and class discussion supports communication and reasoning:

- If we draw a student's name out of a box, will it more likely be a student whose family had one pet or two pets?
- If we draw a student's name out of a box, what numbers of pets for his or her family will be least likely?
- If we draw a student's name out of a box, will it more likely be a student whose family has a pet or a student whose family has no pets?
- If a new student joins the class, what would you guess for the number of pets in his or her family?
- Reinhold and his classmates attend a school of 300 students. From the information in table 6.1, predict the number of students in the school with exactly one family pet.

Students can also create visual representations of the function. Figure 6.1 shows three ways to visualize the family pet data in table 6.1, thereby

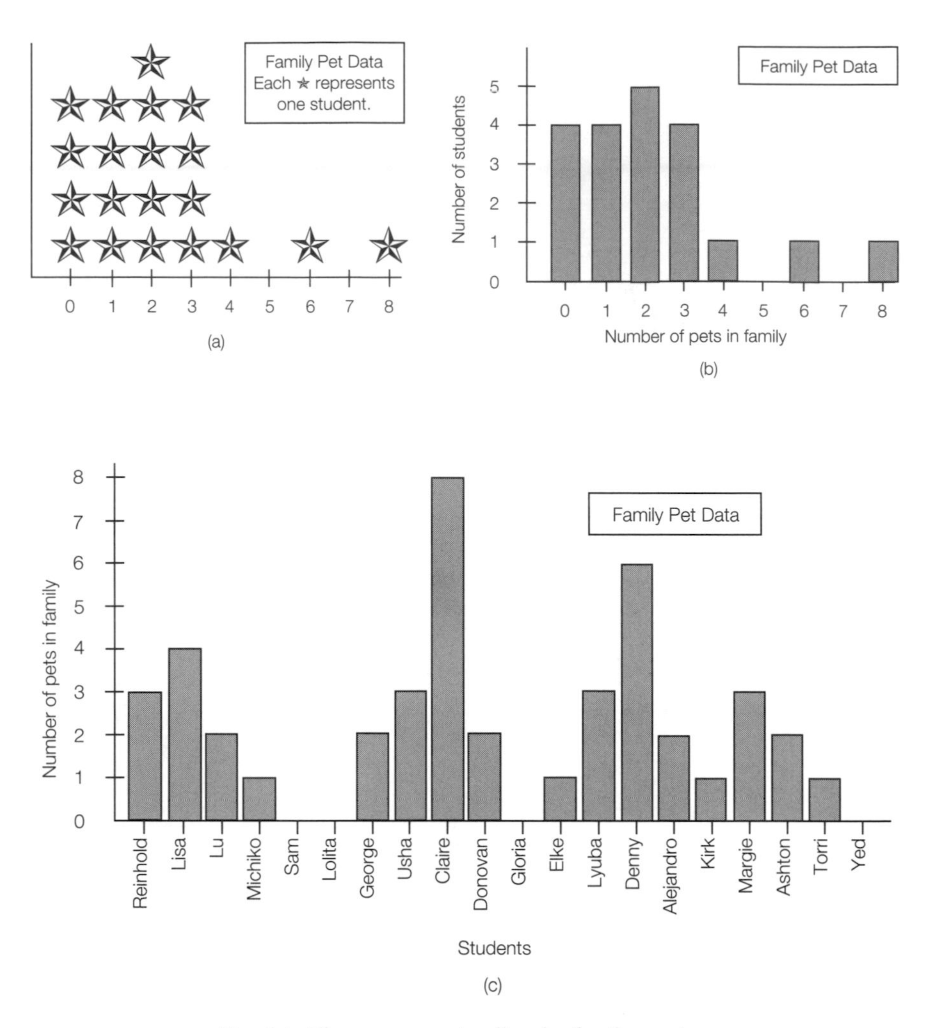

Fig. 6.1. Three ways to visualize the family pet data

connecting the functional relationship to data analysis and statistics. These are just a few of the mathematical connections we can make while exploring functions in the primary grades.

Middle Grades

In the middle grades, students' investigations of the length, width, perimeter, and area of rectangles offer fruitful settings within which functional relationships emerge and through which mathematical connections can be made. Measurements collected by the students, like those in figure 6.2 organized in a spreadsheet for efficient and accurate calculations, present a context for discussing the function concept. A specified length and width produce one unique perimeter and one unique area. The teacher can discuss with students whether a specified length and width yield more than one unique perimeter or area, and whether a specified area or perimeter requires a unique length and width.

	A	B	C	D
1	Length	Width	Perimeter	Area
2	4	5	18	20
3	9	5	28	45
4	3	2	10	6
5	4.5	6	21	27
6	3.78	2	11.56	7.56
7	40	4.5	89	180
8	7	7	28	49
9	4	10	28	40
10	1	6	14	6

Fig. 6.2

When students investigate measurements associated with rectangles, they expand their notion of function in two important ways:

1. To determine either perimeter or area, two inputs are required: length and width. Much later in students' mathematics careers, perhaps when they are studying calculus or linear algebra, instructors routinely assume that students are familiar with multivariable functions and with such descriptive symbolism as $f(x_1, x_2) = x_1 \cdot x_2$ or $p(l, w) = 2l + 2w$. With examples like area and perimeter, teachers help broaden students' conceptions so that they see multivariable relationships as variations of a familiar friend, the function.

2. The area concept helps students see functions as more than simply linear relationships. The classic problem of maximizing the enclosed area of a rectangular-shaped animal pen having a specified perimeter provides a familiar practical setting for students:

> With 100 meters of fencing available to build a temporary holding pen for sheep, what length and width will result in a rectangle with the greatest area?

Figure 6.3 shows lengths and widths of possible holding pens for the fixed perimeter of 100 meters. From a scatterplot of rectangle length and

	A	B	C
1	Length	Width	Area
2	1	49	49
3	5	45	225
4	6.4	43.6	279.04
5	10	40	400
6	15	35	525
7	18.5	31.5	582.75
8	20	30	600
9	25	25	625
10	30	20	600
11	35	15	525
12	40	10	400
13	42.1	7.9	332.59
14	45	5	225
15	50	0	0

Fig. 6.3. Lengths, widths, and areas of rectangles with a perimeter of 100 meters

associated area generated with a spreadsheet, graphing calculator, or appropriate software (fig. 6.4), students observe a set of points that do not follow a straight line pattern. After recognizing and discussing the continuous nature of the measurements, students agree that there is justification for connecting the plotted points to create a smooth curve (fig. 6.5).

The length-area function for rectangles furnishes a gateway to several mathematical connections. Numbers (fig. 6.3), graphs (figs. 6.4 and 6.5), and symbols (fig. 6.6) offer multiple representations to describe how rectangle length corresponds to rectangle area when the perimeter is fixed. Within this exploration, students can discuss the properties of rectangles. Geometric properties of squares and rectangles will likely be compared, and some students will undoubtedly claim that a square is not a rectangle!

Functions embedded in familiar contexts invite the development of middle-grade students' analytical skills. Foreshadowing more formal study of functions, the problem setting described invites questions that develop

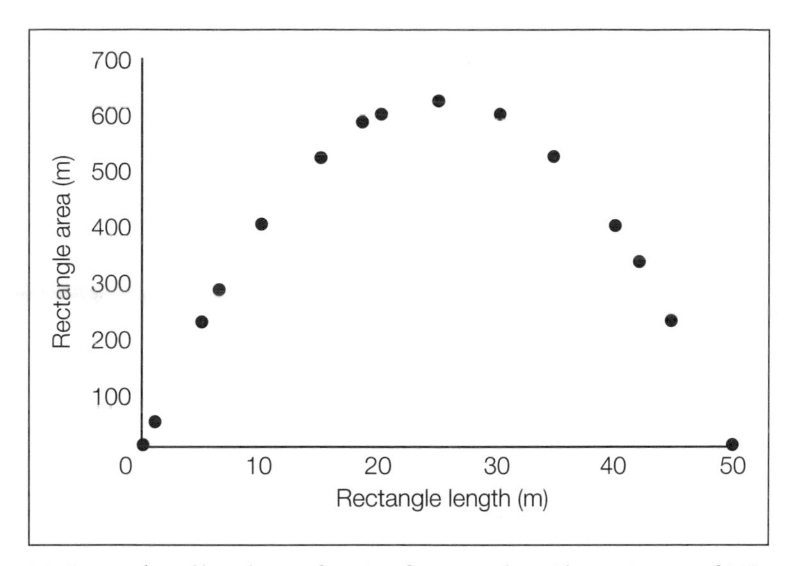

Fig. 6.4. Scatterplot of length-area function for rectangles with a perimeter of 100 meters

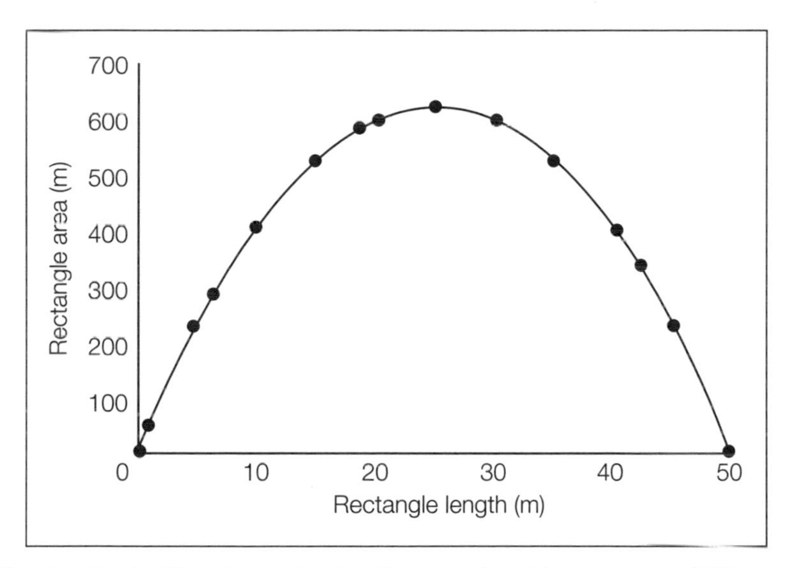

Fig. 6.5. Graph of length-area function for rectangles with a perimeter of 100 meters

students' intuitive notions of domain, range, maximum and minimum values of a function, and limits:

- What is the longest length a rectangular animal pen could have (given a specific perimeter/area)? What is the shortest length it could have?
- What is the largest area that could be enclosed by the animal pen? The smallest area?

- What size of pen generates the largest area? The smallest?

Students can justify their responses by referring to one or more of the length-area representations they have generated.

L: rectangle length W: rectangle width P: rectangle perimeter A: rectangle area	$P = 2L + 2W$ $A = LW$ $P = 100$	$\Rightarrow 100 = 2L + 2W$ $\Rightarrow W = 50 - L$ $\Rightarrow A = L \bullet (50 - L)$ $\Rightarrow A = 50L - L^2$

Fig. 6.6. Symbolic representation of length-area relationship

High School

The next example has been used with secondary school students to investigate and visualize another nonlinear function. The problem below describes a discrete rather than a continuous relationship, an extension important for students but one that is often overlooked in the curriculum. The problem describes a real-world incident; it makes a connection to the physical sciences and offers an opportunity for mathematical modeling.

> While living and working in Norway, I drove a 1967 Volkswagen "beetle." As cold weather approached, the mechanic at a local garage suggested I use a gasoline additive to absorb any moisture that might condense in the fuel line. I dutifully purchased the product, referred to generically as *deicer,* and poured it in with the gasoline when I next filled the tank.
>
> The information on the deicer label suggested regular use during cold weather and recommended adding one can with every fuel fill. I made a mental note to do that, but at the same time I began to wonder: *What happens in that fuel tank as I add more and more deicer? Won't the tank eventually be filled with just deicer?*

Several functions can be identified in the situation (fig. 6.7). Here we examine one of them, the amount of deicer in the gasoline tank as the tank is repeatedly filled. To explore this relationship, and others that can be drawn from the situation, we need to identify the quantities that may vary, describe mathematically how the variables are related, and establish initial conditions and assumptions. A class discussion addressing these concerns affords an ideal setting for introducing or reinforcing mathematical modeling. The questions below will likely emerge from a discussion of the situation:

- How much gasoline does the fuel tank hold?
- How much deicer is added?
- How often is deicer added?
- Is the gasoline tank empty when deicer is added? If not, how full is it?

- What happens to the deicer in the tank? Does it mix with the gasoline? Does it stay in the tank? Does it burn off at the same rate as the gasoline?

- Total amount of deicer in fuel tank after each fill-up
- Increase in the total amount of deicer in fuel tank after each fill-up
- Amount of gasoline in fuel tank after each fill-up
- Decrease in the amount of gasoline in fuel tank after each fill-up
- Ratio of amount of deicer in tank to amount of gasoline in tank after each fill-up
- Ratio of amount of deicer in fuel tank to tank capacity after each fill-up
- Ratio of amount of gasoline in fuel tank to tank capacity after each fill-up

Fig. 6.7. Some functional relationships emerging from the deicer situation

When we explore real-world relationships and try to represent them with mathematical models, we often need to make certain assumptions, accept specific constraints, and establish methods of representation. Students' questions similar to those above help emphasize the need for assumptions, constraints, and representations. From a class discussion, students may come to agree to the following assumptions:

- When combined with gasoline in the fuel tank, deicer mixes uniformly with the gasoline and burns off at a rate equal to that of the gasoline.
- The same amount of deicer is added with each fuel fill-up.
- Fill-ups occur at the same point of "tank emptiness" each time, and the tank is filled to capacity each time it is filled.

To establish initial conditions, it is necessary to stipulate values for tank capacity, amount of deicer added with each fill-up, and the point of tank emptiness at which the tank is refilled. For example, suppose that one quart of deicer is added to a car with a 10-gallon fuel tank every time the fuel tank is half full.

Figure 6.8 shows how a student using a spreadsheet might explore the problem situation with those initial conditions. A functional relationship between elements of columns A and C of the spreadsheet describes the amount of deicer in the gasoline tank as the tank is repeatedly filled. Ordered pairs of the form (fill number, deicer in tank) can be extracted from the spreadsheet: (1, 1), (2, 1.5), (3, 1.75), (4, 1.875), (5, 1.9375), and (6, 1.96875). A scatterplot (fig. 6.9) gives a graphical representation of the function.

Class discussion again helps reinforce the concept of function. For each particular fill-up there is a unique amount of deicer already in the fuel tank. Any of the representations—spreadsheet table, ordered pairs, scatterplot—can be used to verify the nonlinear nature of the functional relationship.

	A	B	C	D
1	After Fill #	Gasoline in Tank (quarts)	Deicer in Tank (quarts)	Increase in Deicer (quarts)
2	1	39	1	1
3	2	38.5	1.5	0.5
4	3	38.25	1.75	0.25
5	4	38.125	1.875	0.125
6	5	38.0625	1.9375	0.0625
7	6	38.03125	1.96875	0.03125
8	7	38.015625	1.984375	0.015625
9	8	38.0078125	1.9921875	0.0078125
10	9	38.00390625	1.99609375	0.00390625

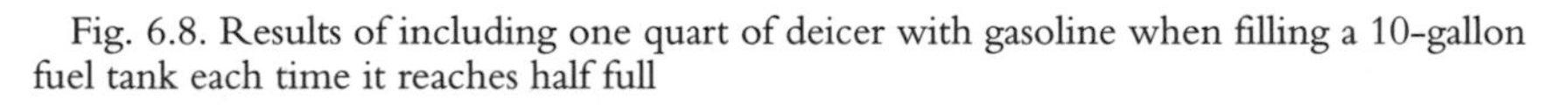

Fig. 6.8. Results of including one quart of deicer with gasoline when filling a 10-gallon fuel tank each time it reaches half full

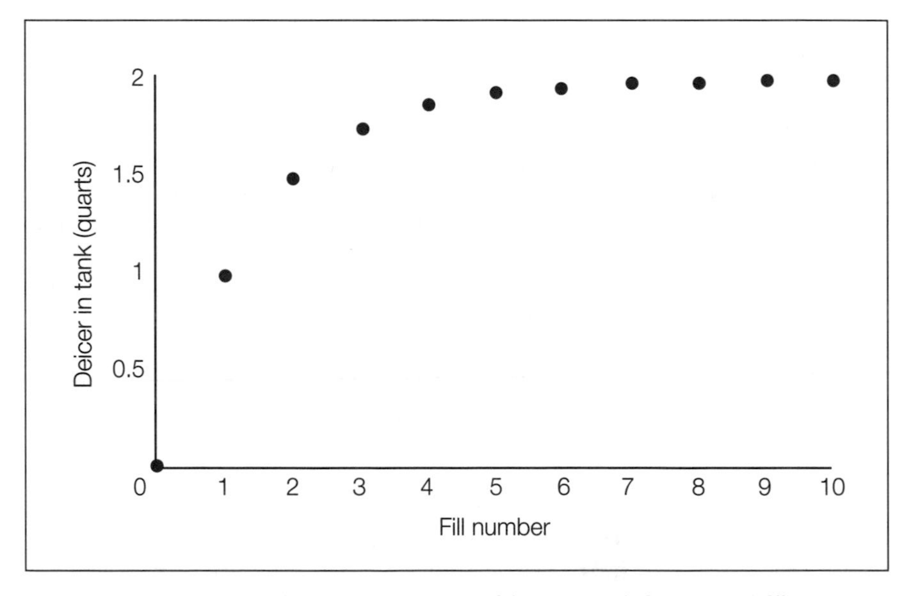

Fig. 6.9. Scatterplot showing amount of deicer in tank for repeated fills

The scatterplot illustrates well the discrete nature of the function (are there elements such as fill-up number 1.4 or fill-up number 5.823?). Although we can explore what happens to the amount of deicer in the tank *between* fills (a continuous relationship), the functional relationship between a particular fill-up and the amount of deicer in the tank at that fill-up is indeed discrete.

With a spreadsheet like the one shown in figure 6.8, students can easily change initial conditions and investigate the impact on the amounts of deicer and gasoline in the tank. Using technology, students can focus on the patterns that emerge from their "What if ...?" questions rather than on the details of arithmetic calculations. Along with the consideration of the multiple representations of these relationships and their connections to mathematical modeling, the patterns that emerge allow the students to make mathematical connections with sequences and series.

For example, the elements in column D of the spreadsheet in figure 6.8 (1, 0.5, 0.125, 0.0625, 0.03125, ...) can be used to introduce or reinforce the concept of *geometric sequences,* whereas the elements in column C (1, 1.5, 1.75, 1.875, 1.9375, 1.96875, ...) are part of a *sequence of partial sums* for the *infinite series* 1 + 0.5 + 0.125 + 0.0625 + ... The connection to sequences and series, in turn, fosters connections to algebraic symbolism as another way to represent the function. For instance, students can concisely represent the elements of the geometric sequence in column D of figure 6.8 as

$$A_n = \left(\frac{1}{2}\right)^{n-1}$$

for $n = 1, 2, 3, \ldots$.

The sequences-and-series connection leads naturally to reasoning and forms of proof as well as to convergence, divergence, and the limit concept. In figure 6.10, symbols are manipulated and the limit concept is applied to abate the original concern about the deicer ultimately dominating the fuel mixture. For the specified initial conditions, the amount of deicer in the tank approaches two quarts. Students may begin to draw the same conclusion by extrapolating from the scatterplot in figure 6.9 or from the pattern in the sequence of values shown in column C of figure 6.8.

CONCLUSIONS

Examples of mathematical connections drawn from functional relationships have been illustrated from three levels of the school mathematics curriculum. Teachers can help students explore functions by collecting data relevant to students' lives and by investigating situations familiar to them. Once generated, relationships can be used to introduce, reinforce, and extend students' conceptions of function. Functional relationships also give rise to a network of mathematical connections. By drawing on

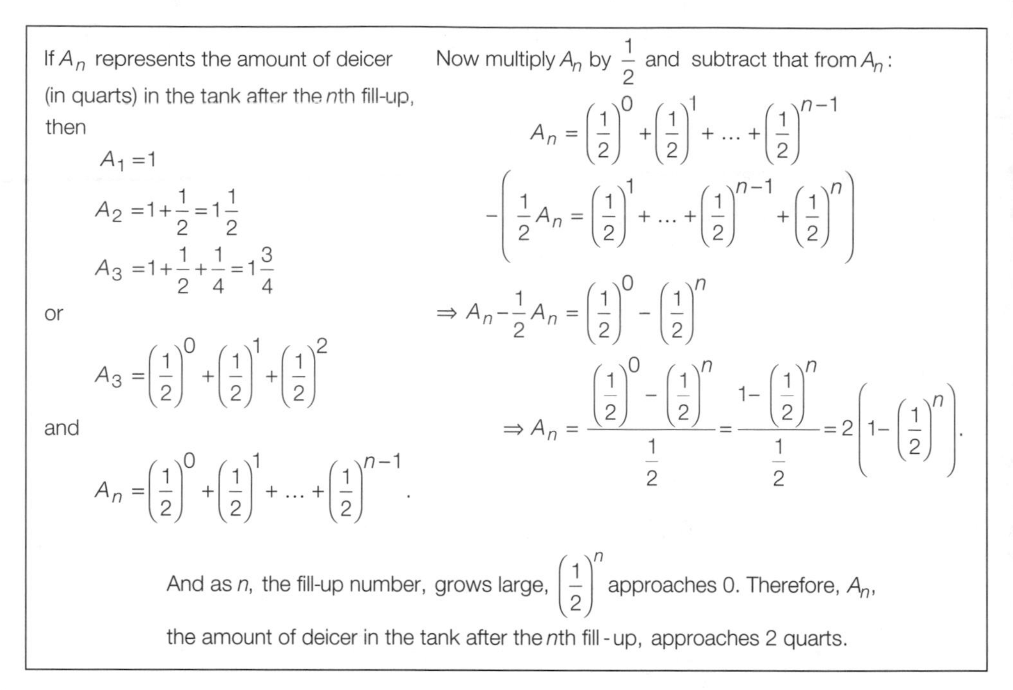

Fig. 6.10. Symbolic argument that the amount of deicer in tank will stabilize

functional relationships to help build bridges among topics of mathematics and the real world, teachers help their students see mathematics as integrated and alive.

REFERENCES

Kleiner, Israel. "Evolution of the Function Concept: A Brief Survey." *College Mathematics Journal* 20 (September 1989): 282–300.

Markovits, Zvia, Bat Sheva Eylon, and Maxim Bruckheimer. "Difficulties Students Have with the Function Concept." In *The Ideas of Algebra, K–12,* 1988 Yearbook of the National Council of Teachers of Mathematics, edited by Arthur F. Coxford, pp. 43–60. Reston, Va.: The Council, 1988.

National Council of Teachers of Mathematics. *Curriculum and Evaluation Standards for School Mathematics.* Reston, Va.: The Council, 1989.

Vinner, Shlomo, and Tommy Dreyfus. "Images and Definitions for the Concept of Function." *Journal for Research in Mathematics Education* 20 (July 1989): 356–66.

7

Making Connections with Transformations in Grades K–8

Rheta N. Rubenstein
Denisse R. Thompson

> We need to construct curricula with greater vertical continuity, to connect the roots of mathematics to the branches of mathematics in the educational experience of children.
>
> —Lynn Arthur Steen, *On the Shoulders of Giants*

WHAT do the following have in common?

- A primary-grade student examines a shell through a magnifying glass.
- An upper elementary school pupil builds a scale model of an airplane.
- A middle school student recognizes that when the sides of a triangle are tripled, nine copies of the original triangle fit inside the enlargement.
- A middle school student notices that if everyone gets a bonus of 10 percent of one's original test scores, the new class average is 10 percent higher than the original average.

Each student described above is experiencing some aspect of the concept of *size change* (sometimes called *dilation*), an enlargement or reduction that produces similar figures. In the examples given, the changes involve magnification, scaling, proportionality, and stretching—topics that students too often see as four separate areas of study. Recognizing that one simple idea can underlie apparently disparate experiences makes learning mathematics easier and more meaningful for students.

The broader topic of transformations (of which size change is one type) can connect many ideas from primary grades through high school and beyond. Many reasons exist for including transformations prominently and consistently in the mathematics curriculum:

1. Accessibility—Slides, flips, turns, and stretches are kinesthetic and easy for students at all levels to manipulate and see.

2. Communication—The language of transformations offers a simple way for students to describe a variety of phenomena.

3. Problem solving—Symmetries, rotations, and proportions in size changes present avenues of solution for many problems.

4. Connections—Transformations support applications of mathematics, such as in design and construction, and unify many ideas within mathematics, including conditions for congruence and similarity, coordinate and synthetic geometry, and measures of data affected by various changes. Seeing the unity of these different topics simplifies students' learning. Too often students think that hundreds of different skills must be learned each year; we serve them well when we minimize the number of disconnected concepts by highlighting such natural pedagogical umbrellas as transformations.

Transformations can be integrated into elementary and middle school programs in ways that support objectives within and beyond mathematics and that lay a groundwork for more advanced topics. In early grades, transformations may be handled informally and integrated with play, art, physical education, literature, and science. Young students can develop an intuitive understanding by visualizing and describing shapes and relationships. Older students can continue their informal work while becoming more analytical through using measurement, ratios, and symbolism in their investigations. These experiences pave the way for more formal and abstract work with transformations in high school and beyond. (See the articles by Crowley and Hirschhorn and Viktora in this yearbook for connections at the high school level.)

PRIMARY GRADES

Many transformation ideas can be introduced in the primary grades during students' first encounters with geometry. For instance, congruence arises naturally in sorting toys and shapes, some of which are identical, and fitting puzzle pieces into their own spaces. Similarity, too, can be informally introduced with nesting shapes (e.g., open cubes that fit inside one another) that students can describe as having the same shape but different sizes. Thus, students begin to focus on the underlying properties and relationships of objects—important aspects of geometric thinking.

Students can be asked to explore and discuss what happens when a given figure is cut into parts and those parts rearranged. *Grandfather Tang's Story* (Tombert 1990) suggests a wonderful connection between literature and mathematics. As the story is read, students see how seven shapes are transformed into several different animals. Given a set of tangrams, students enjoy making the different animals by matching their own tangram pieces to the arrangements in the book. Notions of congruence and spatial sense are nurtured as children turn and slide the pieces until the animals emerge.

Primary school students can use their bodies to act out fundamental transformations such as flips, slides, and turns. As the NCTM *Second-Grade Book* (Addenda series) (Burton et al. 1992) illustrates, students can lie on the floor and flip about their right side, left side, or feet. They can observe one another (or partners) and discuss how their bodies are the same and different before and after the flips. Students observe informally and concretely that the orientation is reversed, yet the original and the final images are the same (congruent). Similarly, students find that a slide across the floor preserves everything except location. Thus, students gain a kinesthetic sense of reflections (flips) and translations (slides). Similar activities can be done with rotations (turns around a point).

After students have used themselves as manipulatives, it is useful for them to explore the same transformations with other objects. Pattern blocks are a versatile medium for such explorations. Consider the activity in figure 7.1,

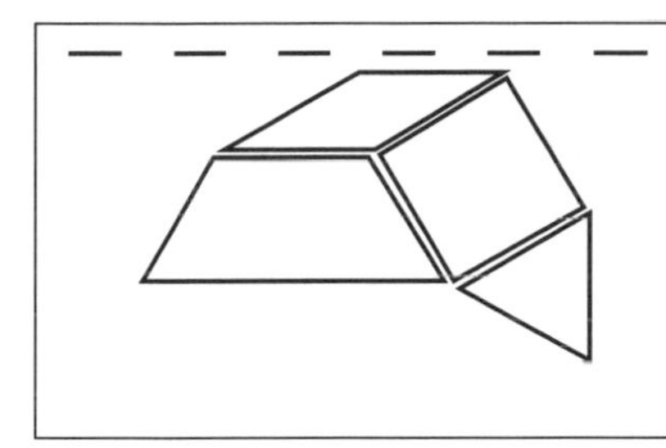

Fold the paper on the dotted line.
Trace the design. Unfold the paper.
Place pattern blocks on both the design and the tracing.

How are they the same?
How are they different?

Fig. 7.1. Reflections with pattern blocks

adapted from Pasternak and Silvey (1975). Here students reflect a pattern across a line. Once again, they compare the original shape and the resulting image. Students can also explore transformations with pattern blocks on the computer. In the software package Exploring Mathematics with Manipulatives (1992), students use transformation ideas to solve a puzzle. They take a screen image of a pattern block and turn, reflect, or slide it until the piece fits into a given slot on the screen. All these activities help construct intuitive understandings that lay a foundation for more-formal studies later.

In addition to slides, flips, and turns, primary-grade students can also explore symmetry. Paper-folding activities like producing valentines and evergreen trees are a natural way to start. Symmetry can also be connected to nature studies when students identify plants and animals that "match on both sides of a line." Figure 7.2 illustrates a cut-and-fold symmetry exploration.

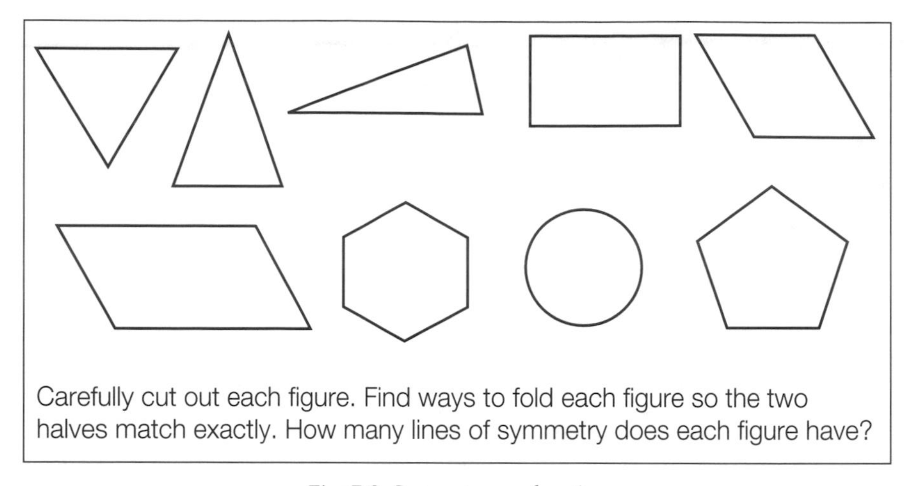

Carefully cut out each figure. Find ways to fold each figure so the two halves match exactly. How many lines of symmetry does each figure have?

Fig. 7.2. Symmetry exploration

After students have folded these designs to find lines of symmetry, they can try constructing their own designs with a given number of lines of symmetry. Again, pattern blocks supply a helpful medium for students to use in identifying and constructing figures with one, two, or four lines of symmetry. Children may create designs as simple or as complex as they like. They can discuss their designs with classmates and use Miras or mirrors to check their conjectured symmetry lines.

Ideas from plane figures can be extended to three-dimensional figures. Students can use clay shapes or Styrofoam models and cutting tools to explore the symmetry of various solids; they can determine if it is possible to slice the figure to make two shapes that are exactly the same. Thus, they begin to see that slicing a solid figure to form congruent halves is the three-dimensional equivalent of folding a two-dimensional figure.

UPPER ELEMENTARY GRADES

Symmetry activities in the upper elementary grades can connect to physical education. Whereas younger students used their bodies to flip, turn, and slide, older children can create movements or dances in which all their actions are symmetric. Students discover that although moving symmetrically means they can jump, flap their arms like wings, and do knee bends, they can't walk, hop on one foot, or wave one hand. With individual partners, however, they can move in more ways, as long as the two people move as mirror images of each other.

Students can be introduced to rotations (turns) through explorations of doorknobs, clock hands, and their own bodies in motion. Students can explore whether a given figure has rotational symmetry by experimenting to see if it can be turned around a point until it matches with itself. (For example, a square can be turned around its center four times, each a 90° turn; a regular hexagon can make six 60° turns around its center.) Figure 7.3 illustrates a rotational-symmetry activity with pattern blocks. After such an activity, students can create designs of their own with half-turn (180°) symmetry, quarter-turn (90°) symmetry, or other angles of turn symmetry. Such an open-ended activity elicits creativity and a variety of "correct" responses.

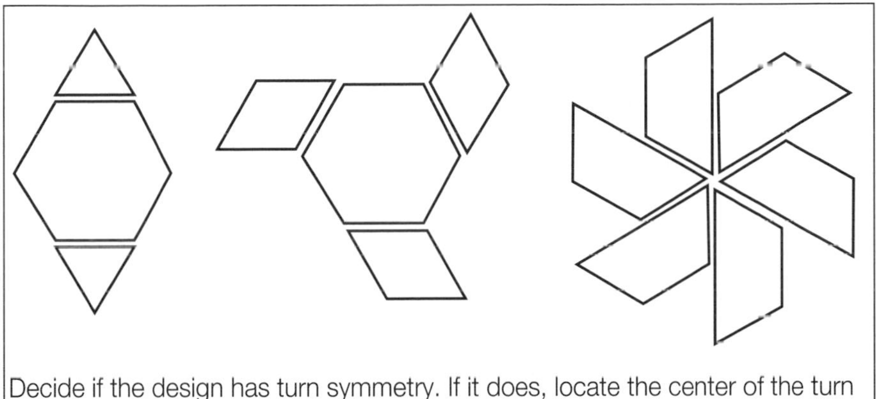

Decide if the design has turn symmetry. If it does, locate the center of the turn and tell how many degrees the design must be turned to first match with itself.

Fig. 7.3. Rotations with pattern blocks

Size changes (stretching and shrinking to produce similar figures) can be introduced with cartoon figures such as those in figure 7.4. These images help build a sense of similarity by illustrating the differences between size changes and distortions, that is, transformations in which stretches are not equal in both dimensions. (The images in figure 7.4 were produced with a computer drawing tool. When a corner of the picture's frame is dragged, a similar figure results; when an edge is dragged, a distortion results.)

Freudenthal (1987, p. 286) relates an interesting sequence of lessons on similarity and its implications: Students enter their classroom one morning and discover that the window is open and on the chalkboard is an enormous handprint. After deciding that a giant must have visited, they wonder how tall it was:

> "Look at my hand." The teacher puts her hand on the giant hand-print, which appears to be four times as [long] as hers. The teacher is measured. A string is cut off to a length four times as great as the teacher's height. The children write a letter to the giant on the blackboard. "This is your height," they write. The next day the giant has answered the letter.

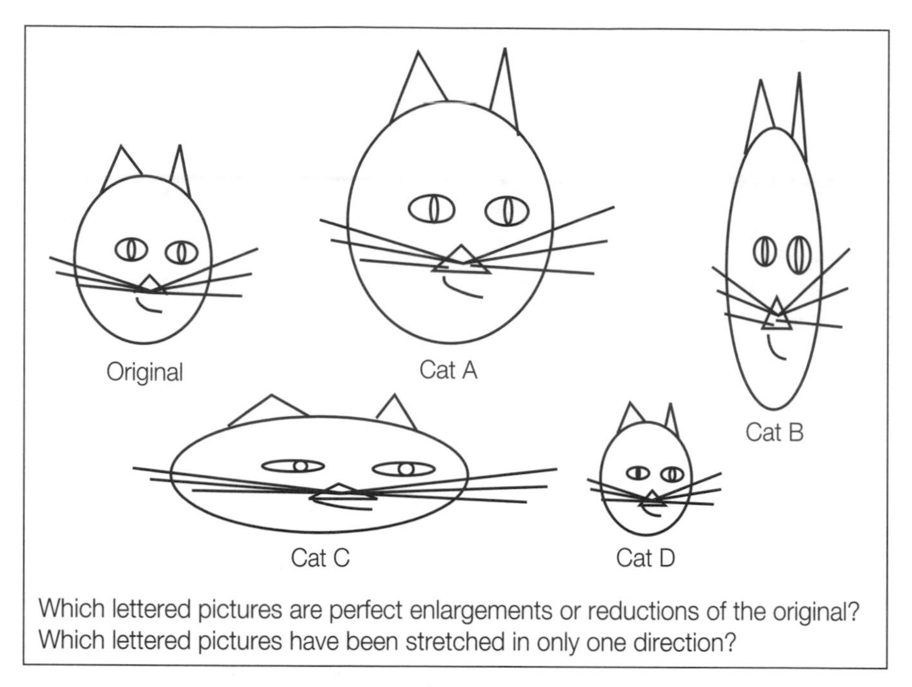

Fig. 7.4. Size changes with cartoons

So begins a series of lessons investigating giant-sized newspapers, mittens, cakes, and the corresponding measures of lengths, areas, and volumes. (For a related activity, see the article by Krulik and Rudnick in this yearbook.)

Del Grande (1989) reported a serendipitous use of transformations in a problem-solving context. In searching for all the squares on a geoboard, a student used four elastic bands to make the display shown in figure 7.5 and wondered if the shape formed was a square. Although the students offered many suggestions about how to answer this question, one student in particular used rotations. By making successive quarter turns about the center of the geoboard, he showed that the design stayed the same. Therefore, all the sides and all the angles must be equal, he reasoned, so the figure must be a square!

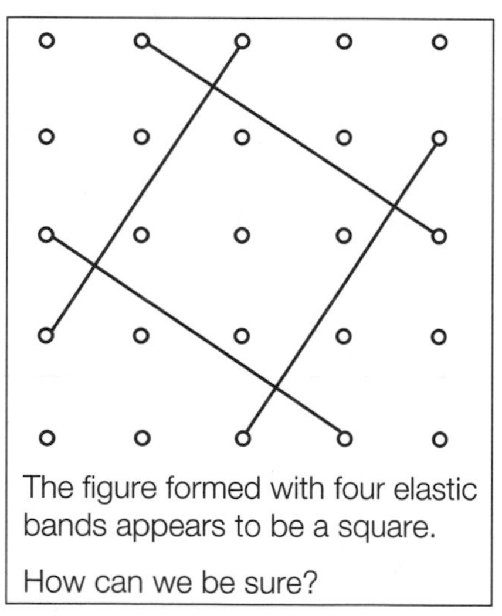

Fig. 7.5. Reasoning on a geoboard

MIDDLE SCHOOL LEVEL

At the middle school level, many of the ideas introduced earlier are deepened and extended. For example, trademarks and company logos can be analyzed for line and rotational symmetry (Renshaw 1986). Students may also be interested to learn that commercial artists perceive trademarks with line symmetry as indicating stability, whereas those trademarks having rotational symmetry indicate movement. Students may want to try creating their own logos with line or rotational symmetry. Kim (1981) offered inspiring examples as well as strategies for creating amazing visuals in which English words have line or rotational symmetry. Figure 7.6 shows our own attempts to apply Kim's techniques.

Rotate WISCONSIN a half-turn around the center of the O to see Wisconsin again.

Find a mirror line through OHIO so the word reflects itself.

Make your own word or design with half-turn or line symmetry.

Fig. 7.6. Symmetry with words

In the middle school, flips (reflections), turns (rotations), and slides (translations) can be used to create the seven possible border patterns shown in figure 7.7, the same patterns found in the artwork of the people of the San Ildefonso Pueblo (fig. 7.8). (See Zaslavsky [1973] for examples from other cultures.)

Notions of reflections can be extended to include bisector and construction activities with Miras. (See Geddes [1992, pp. 66–67]). The composition of transformations can also be foreshadowed. When students reflect a figure over a line, then reflect the image over a line parallel to the first, they notice that the final image is a translation of the original. Likewise, students discover that the same experiment done with intersecting lines produces a rotation of the original figure. These activities build foundations for later work with the composition of functions. (See the article by Crowley in this yearbook.)

Students can further explore enlargements, reductions, and similarity. One activity that students enjoy is to copy figures from a small-scale grid to a larger grid (see fig. 7.9). The Middle Grades Mathematics Project unit on similarity (Lappan et al. 1986) offers several other ways for students to explore size changes, including using elastic-band stretchers, point projections, and coordinates (see fig. 7.10a–c). Students learn that in size changes, angle measures are preserved and corresponding lengths are proportional. They

Use graph paper. Make a strip of several congruent rectangular frames. For each border pattern draw a simple shape in the first frame.

Copy your shape repeatedly as the directions and illustrations show.

Dashed lines show reflection lines and dots show centers of half-turns.

Type 1:
Slide into the next frame.

Type 2:
Flip over the horizontal line and slide to the next frame.

Type 3:
Flip over the vertical line into the next frame.

Type 4:
Make a half-turn into the next frame.

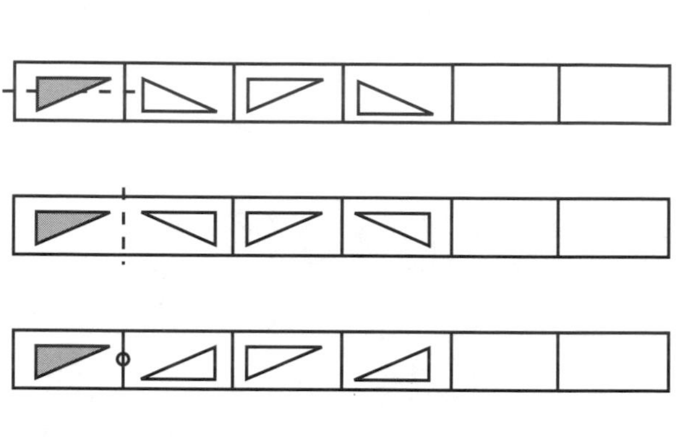

Type 5:
Flip over the vertical line into the next frame. Then, make a half-turn into the following frame. Repeat the flips and half-turns.

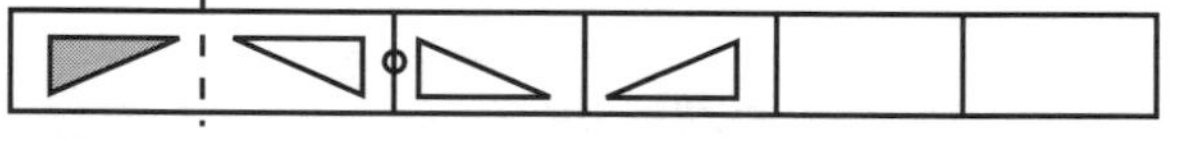

Type 6:
Use a double set of frames. Flip over the horizontal line into the frame below. Then copy the pairs repeatedly.

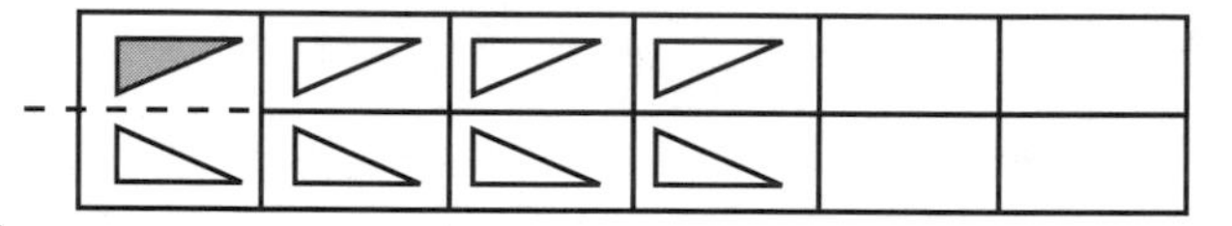

Type 7:
Use a double set of frames. Flip over the horizontal line into the frame below. Then flip the pair over the vertical line into the next pair of frames. Repeat the flips over the vertical lines.

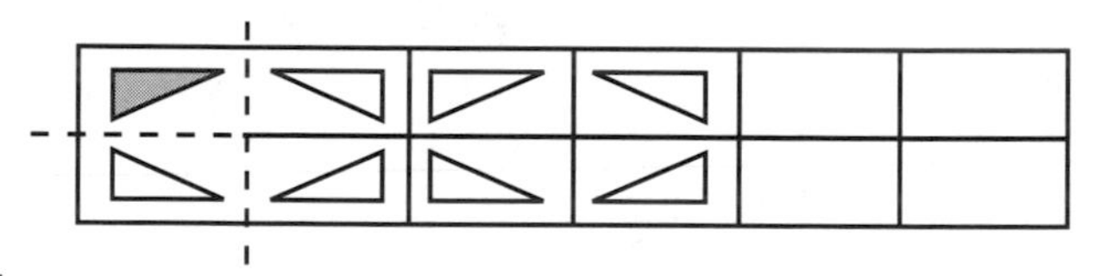

Fig. 7.7. Making border patterns

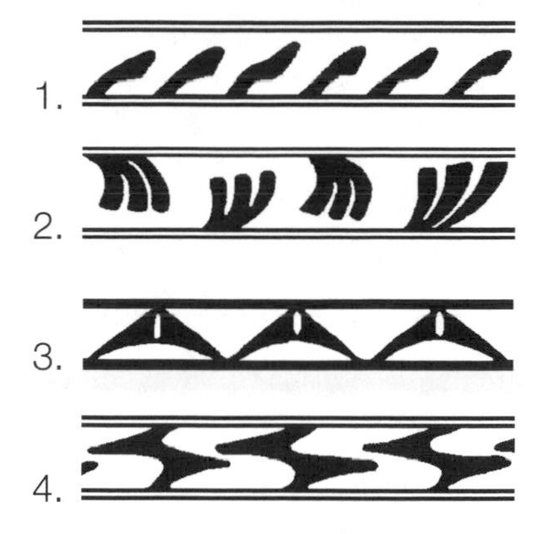

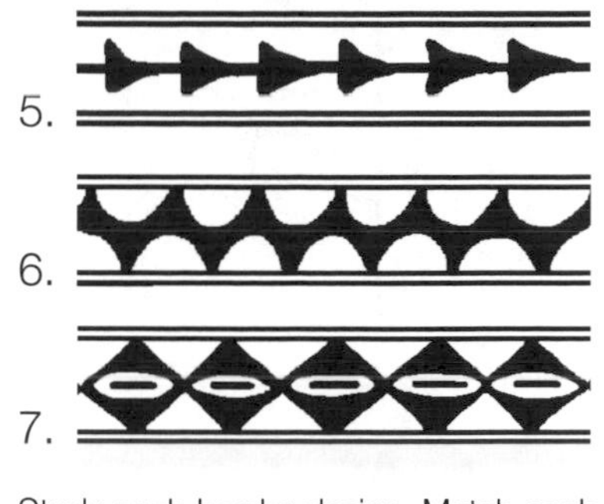

Study each border design. Match each design with a type from figure 7.7. [Key: 1, 2, 3, 4, 6, 5, 7]

Fig. 7.8. Border designs from the pottery of the San Ildefonso Pueblo, New Mexico (Source: Crowe and Thompson 1987, p. 108)

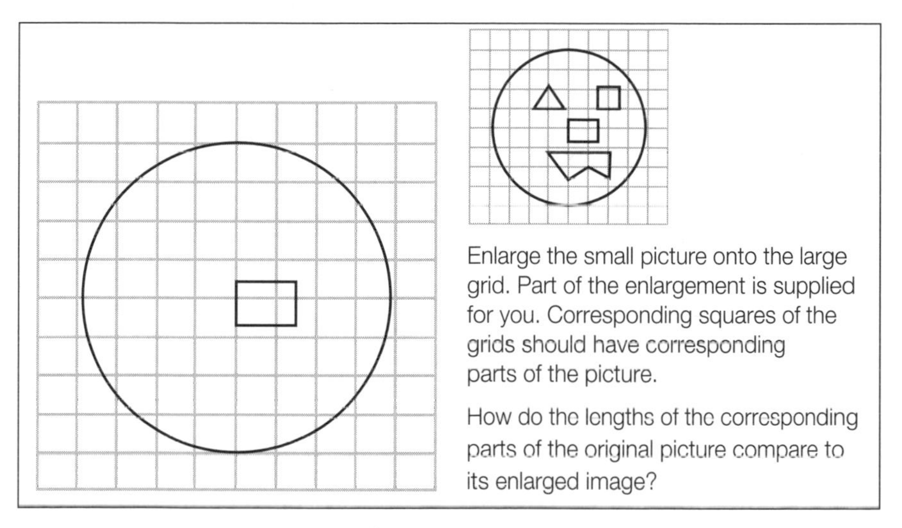

Fig. 7.9. Using different-sized grids to enlarge figures

should be encouraged to recognize enlargement, reduction, and similarity in everyday life, for example, different-sized flags, overhead-projector images, measuring cups, and different-sized nuts and bolts.

Students can also explore areas under enlargements. As shown in figure 7.11, students enjoy finding the n^2 copies of the original triangle that fit inside an image with sides n times the original. They see that areas grow with the square of the magnification factor. When students attempt to "undo" enlargements ("How do you shrink the larger figure down to the smaller?"), they deepen their understanding of reciprocals and begin to learn about inverses.

Further, students can look for instances of congruence in everyday life (for example, keys to the same lock, school desks, and interchangeable parts

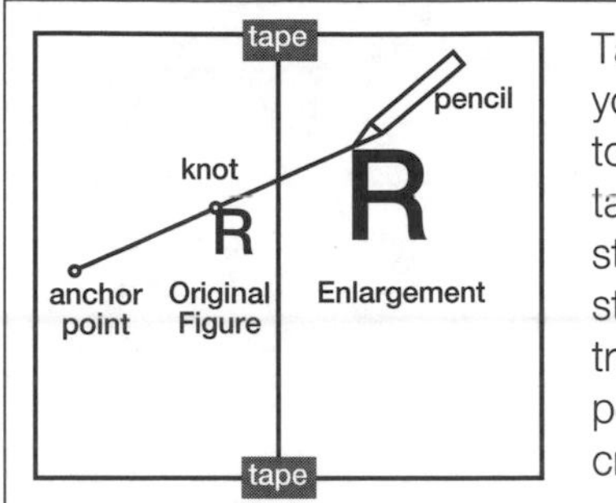

Tape an original picture and a blank paper to your desk. Interlace two equal elastic bands to form a knot; this is your stretcher. With a tack or your thumb, anchor one end of the stretcher to the original paper. Pull on the stretcher with the pencil tip so the knot traces the original figure. As you do so, the pencil point touching the blank paper will create the enlargement.

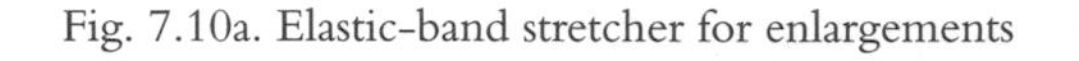
Fig. 7.10a. Elastic-band stretcher for enlargements

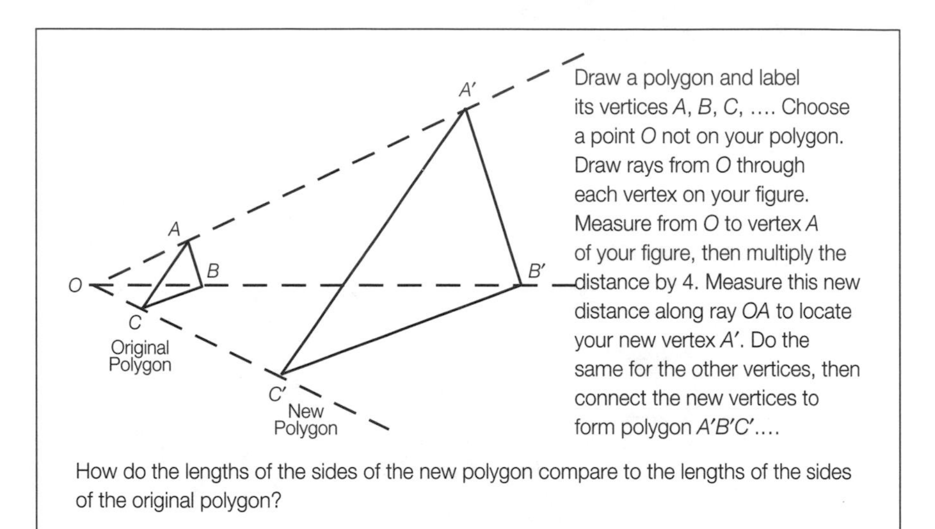

Fig. 7.10b. Point projection by a scale factor of 4

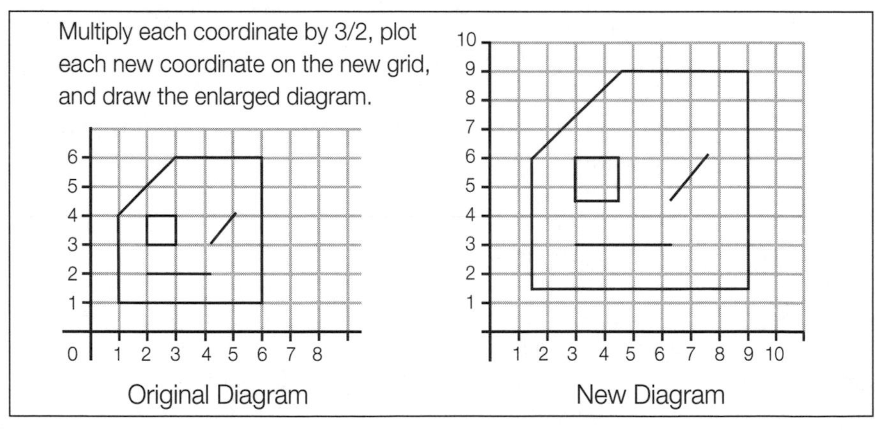

Fig. 7.10c. Size change with coordinates

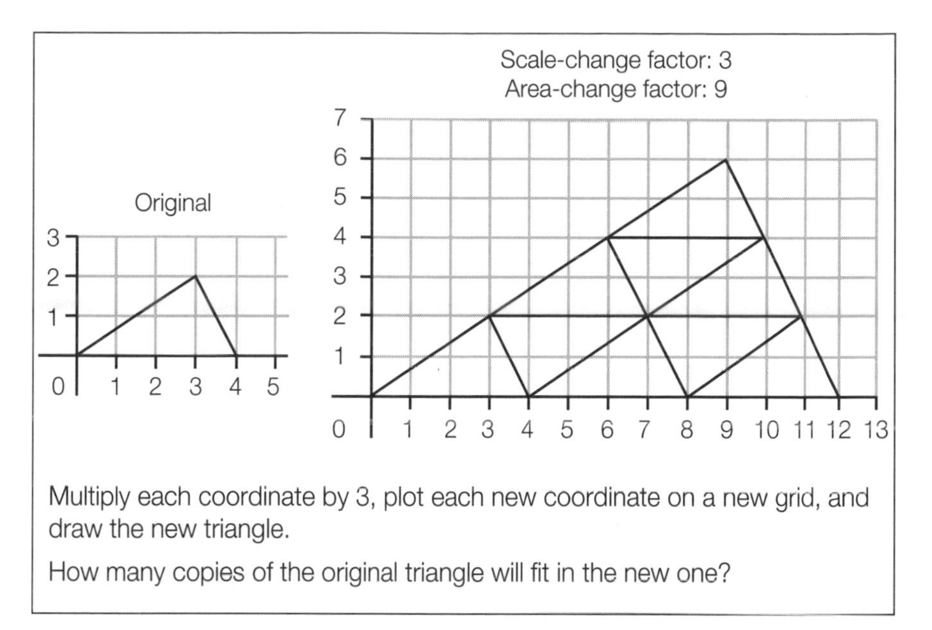

Fig. 7.11. Relating size changes to area changes

in manufactured goods) and connect them with slides, flips, and turns (but not with enlargements). They can recognize that slides, flips, and turns produce congruent figures, but size changes (shrinking or enlarging) are needed for similarities that are not congruent. Notions of congruence and similarity developed when using transformation activities are much broader than the traditional definitions that students typically encounter. First, these new notions of congruence and similarity are not restricted to polygons or even two-dimensional figures. Second, the inclusive relationship is highlighted: congruence is a special instance of similarity where the "stretch" factor is equal to 1. Formal proofs of these concepts are facilitated in high school by the informal work accomplished in middle school.

By grade 8, students begin to synthesize, generalize, and formalize many ideas begun earlier, and they can integrate concepts from algebra with those from geometry. In the article that follows, Crowley describes activities for extending transformations to the coordinate plane, specifically to include translations, reflections over axes, rotations of 90 degrees, and size changes.

When exploring transformations, teachers can initiate significant discussions by asking, "What stays the same and what changes? And if something changes, can we predict how?" Again, students integrate geometry and algebra when they discover that reflections, rotations, and translations preserve lengths; translations preserve slope; and size changes preserve ratios of lengths within figures. Rotations of 90 degrees change slopes, but, predictably, each segment is perpendicular to its original.

Asking what stays the same and what changes applies to other work with transformations as well. For example, suppose everyone in class gets 5 extra points added to the last test score. *What stays the same and what changes?* The difference between the highest and lowest scores (range) stays the same. The mean changes, but in a predictable way: it is the original mean plus 5 points. Figure 7.12 shows that the graph of the scores moves five units to the right.

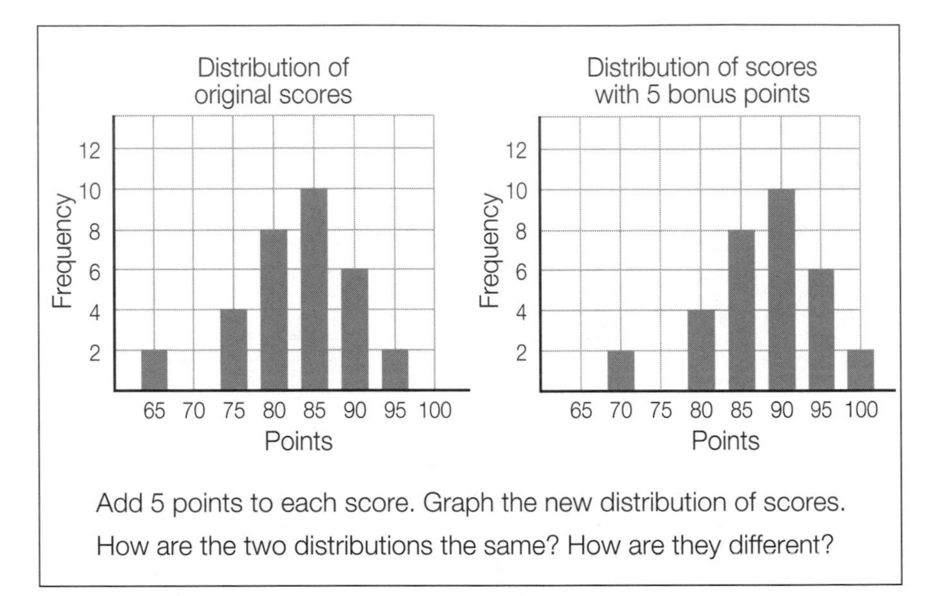

Fig. 7.12. Sliding data

Similarly, students can investigate size changes of data (the results when data are multiplied by a scalar). Figure 7.13 shows a box plot of the distribution of wages of workers in a business if everyone receives a 25 percent pay raise. What stays the same and what changes? In this instance, we are multiplying each value by a factor of 1.25, and although we can't actually use them, we can recall our elastic bands (see fig. 7.10a) and imagine stretching the original box plot to find the new one. Every value is pulled away from the origin by a factor of 1.25. We recognize that the whole picture, including range, median, and percentiles, is increased by 25 percent. Using statistical software or graphing calculators, students can verify these relationships for other data sets. (In more advanced courses, students can prove these phenomena.)

Students see that the concepts of translation and size change span statistics as well as geometry. These ideas also foreshadow an idea in later statistical work where translation and size change are precisely the transformations used to produce standardized test scores. (See the article by Hirschhorn and Viktora later in this yearbook.)

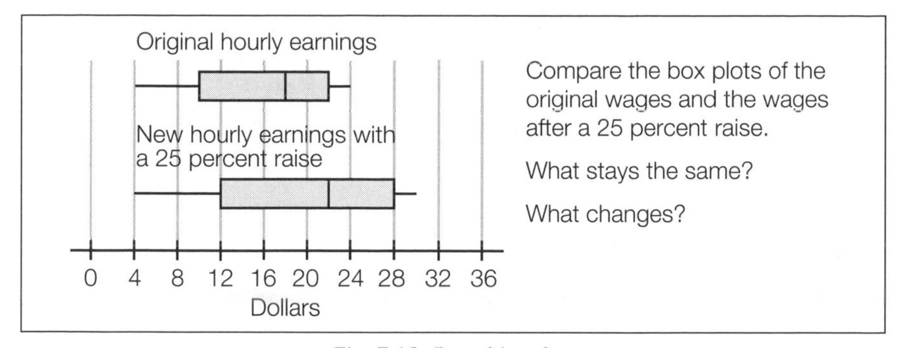

Fig. 7.13. Stretching data

Conclusion

Too often, students see mathematical topics in isolation and fail to see how a topic studied in one domain can be applied to understanding in another domain. Transformations eliminate much of this isolation. As these examples suggest, transformation ideas have the potential to pervade the mathematics curriculum and provide a powerful force to connect and support work in patterning, reasoning, problem solving, spatial sense, algebra, geometry, and statistics. Indeed, "[transforming] is a major theme of contemporary mathematics because it provides a useful and illuminating way to organize relations among shapes and patterns" (Senechal 1990, p. 168). By building transformation ideas into the K–8 curriculum, we also lay the foundation for important ideas to be studied formally at the high school level and beyond. In this way, we are constructing the vertical continuity that Steen advocated in the opening quotation.

References

Burton, Grace, et al. *Second-Grade Book. Curriculum and Evaluation Standards for School Mathematics* Addenda Series, Grades K–6, edited by Miriam A. Leiva. Reston, Va.: National Council of Teachers of Mathematics, 1992.

Crowe, Donald W., and Thomas M. Thompson. "Some Modern Uses of Geometry." In *Learning and Teaching Geometry, K–12,* 1987 Yearbook of the National Council of Teachers of Mathematics, edited by Mary Montgomery Lindquist, pp. 101–12. Reston, Va.: The Council, 1987.

Del Grande, John. Remarks in address at conference of Southwest Ontario Association of Mathematics Educators, Windsor, Ont., November 1989.

Exploring Mathematics with Manipulatives, Level II. Educational Development Center, Newton, Mass.: 1992.

Freudenthal, Hans. "Mathematics Starting and Staying in Reality." In *Developments in School Mathematics Education around the World,* edited by Izaak Wirszup and Robert Streit. Reston, Va.: National Council of Teachers of Mathematics, 1987.

Geddes, Dorothy. *Geometry in the Middle Grades. Curriculum and Evaluation Standards for School Mathematics* Addenda Series, Grades 5–8, edited by Frances R. Curcio. Reston, Va.: National Council of Teachers of Mathematics, 1992.

Kim, Scott. *Inversions.* Peterborough, N.H.: BYTE Books, 1981.

Lappan, Glenda, William Fitzgerald, Mary Jean Winter, and Elizabeth Phillips. *Similarity and Equivalent Fractions.* Menlo Park, Calif.: Addison-Wesley Publishing Co., 1986.

Pasternak, Marian, and Linda Silvey. *Pattern Blocks Activities A.* Palo Alto, Calif.: Creative Publications, 1975.

Renshaw, Barbara S. "Symmetry the Trademark Way." *Arithmetic Teacher* 34 (September 1986): 6–12.

Senechal, Marjorie. "Shape." In *On the Shoulders of Giants: New Approaches to Numeracy,* edited by Lynn Arthur Steen. Washington, D.C.: National Academy Press, 1990.

Tombert, Ann. *Grandfather Tang's Story.* New York: Crown Publishers, 1990.

Zaslavsky, Claudia. *Africa Counts: Number and Pattern in African Culture.* Westport, Conn.: Lawrence Hill & Co., 1973.

8

Transformations: Making Connections in High School Mathematics

Mary L. Crowley

> The study of transformations provides a fresh insight into standard geometric problems.... But, more important, the transformation point of view serves to unify mathematics. The concept of a transformation illuminates the studies of functions, vectors, groups, matrices, complex numbers and linear algebra.
>
> —Richard C. Brown, *Transformational Geometry*

TRANSFORMATIONS deserve a good spin doctor. They appeal to the intuition. They lend themselves to use with concrete materials. They are easily and effectively studied using graphing calculators and computers. They serve as powerful problem-solving tools. They link traditionally compartmentalized areas of mathematics. And they have applications in areas outside of mathematics. All these characteristics promote the learning of mathematics!

This article focuses primarily on how transformations can be used to make connections among mathematical topics. In so doing, however, many of the other attributes listed above also emerge.

MULTIPLE REPRESENTATIONS

Transformations can be described using a variety of representations. Pictorial depictions such as those in figure 8.1 provide visual representations of the motion associated with each relationship. These models are closely aligned with students' intuition and experiences; flipping, turning, and sliding are activities that students can easily perform and to which they can easily relate. Alternatively, coordinate descriptions—mapping rules, matrices, and vectors—provide numerical or algebraic representations for the transformational tools. For example, when the coordinates of the point (x, y) are written as the column matrix

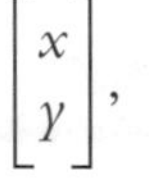

$$\begin{bmatrix} x \\ y \end{bmatrix},$$

the reflection of (x, y) in the x-axis is represented by the mapping rule $(x, y) \rightarrow (x, -y)$ and by the matrix product

$$\begin{bmatrix} 1 & 0 \\ 0 & -1 \end{bmatrix} \cdot \begin{bmatrix} x \\ y \end{bmatrix} = \begin{bmatrix} x \\ -y \end{bmatrix}.$$

These symbolic notations, as we shall see, facilitate the application of transformational concepts.

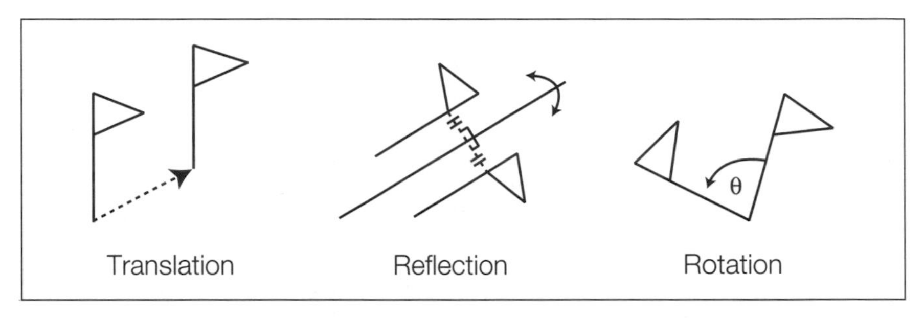

Fig. 8.1. Pictorial representations

Mapping rules and, subsequently, matrix representations can be easily developed by students. To start, ask students to generate coordinates for corresponding preimage and image points using graph paper. For example, for the reflection over the x-axis, the image of (3, 4) is (3, –4); the image of (–5, –1) is (–5, 1); and so on. (Keeping track of these in a table is most useful.) After suitable exploration, students should make conjectures about any general relationships that they see between the coordinates of the corresponding points. These conjectures should be verbalized as well as expressed in expository writing (e.g., "Whenever we flip over the x-axis, it looks like the first number in the pair stays the same but the second number changes its sign."). In turn, responses like this can be formalized into a mapping rule, in this example, $(x, y) \rightarrow (x, -y)$. Activities that promote exploring, conjecturing, and writing about mathematics are important to every student's mathematical development.

PLANE GEOMETRY

Conventionally, transformations are introduced in conjunction with the study of plane geometry. In this context, a transformation is defined to be a one-to-one mapping (a rule, a function) that assigns to each point in the plane some (other) point. The transformations most commonly studied are of two types: those that preserve distance and those that change scale. The distance-preserving transformations, of which there are only four types, are called the *isometries.* They are the reflection in a line, the rotation around a point, the translation, and the glide reflection. The dilation and the stretches are the most commonly studied scale-changing transformations.

With this assortment of tools, the standard notions of Euclidean geometry—congruence, similarity, parallelism, perpendicularity, symmetry, and so on—can be developed. As an example, the notion of congruence can be defined using transformations (e.g., *two shapes are congruent if one is the image of the other under an isometry*). This definition has several advantages. First, it applies to *all* shapes. Thus, unlike in the more traditional approach, it is not necessary to establish one set of conditions for the congruence of triangles and another set of conditions for the congruence of circles, for example. Second, unlike the hand waving about "moving and flipping" that often accompanies the traditional Euclidean definition of congruence, transformations make these types of moves legitimate.

MATRICES

Matrices are one of the topics recommended for increased attention at the grades 9–12 level by the *Curriculum and Evaluation Standards for School Mathematics* (National Council of Teachers of Mathematics 1989). Because most of the special-case transformations that are studied—for example, rotation around the origin or reflection in the line $y = mx$—are linear mappings, they can be represented using matrices. Thus, studying transformations offers an opportunity to reinforce the versatility of the matrix concept and to practice operations with matrices. A knowledge of matrices and their operations can generate specific transformational matrices; transformational images can be found by applying corresponding matrices and operations.

As an introduction to matrix representations, students familiar with matrix multiplication can be given a selected matrix and asked to determine what effect it has on a curve. For example, ask students to let the matrix

$$\begin{bmatrix} 1 & 0 \\ 0 & -1 \end{bmatrix}$$

act on the coordinates of several points on a parabola, say, $y = x^2$. As an example,

$$\begin{bmatrix} 1 & 0 \\ 0 & -1 \end{bmatrix} \cdot \begin{bmatrix} -2 \\ 4 \end{bmatrix} = \begin{bmatrix} -2 \\ -4 \end{bmatrix}.$$

The set of resulting coordinates can then be plotted and the graph of the new curve compared to the graph of the original curve. (A powerful calculator such as the TI-82 or software such as IBM's Mathematics Exploration Toolkit can simplify the multiplications.) This type of exploration also lends itself well to students working in pairs. As one student multiplies, the other student plots the points. When enough points have been plotted, students should make conjectures about which transformation this matrix represents and, using several other curves, establish empirically the validity of their hypotheses. A list of matrix and algebraic representations of the standard transformations is presented in table 8.1.

The question of which transformations can be represented by a two-by-two matrix is an interesting one, and the answer is within the reach of most high school students. Let that matrix be

$$\begin{bmatrix} a & b \\ c & d \end{bmatrix}.$$

From the rules of matrix multiplication, regardless of what a, b, c, and d are,

$$\begin{bmatrix} a & b \\ c & d \end{bmatrix} \cdot \begin{bmatrix} 0 \\ 0 \end{bmatrix}$$

must equal

$$\begin{bmatrix} 0 \\ 0 \end{bmatrix}.$$

Thus, the only candidates for representation by a two-by-two matrix are the transformations that map the origin to the origin. (This is, of course, a characteristic of all linear transformations.) The translation is therefore excluded because, except for the trivial case of $(x, y) \rightarrow (x, y)$, the origin never maps to itself. Rotations around the origin, reflections through a line of the form $y = mx$, dilations with center $(0, 0)$, and stretches of the form $(x, y) \rightarrow (ax, by)$ where $a \neq 0$ and $b \neq 0$ do map the origin to itself and, indeed, are linear. Thus, they can be represented with matrices.

The next question, then, is, If the matrix exists, how can it be determined? As demonstrated below, this is answered if we know the images of the points (1, 0) and (0, 1) for that transformation.

Let T represent a transformation, where $T(1, 0) = (x_1, y_1)$ and $T(0, 1) = (x_2, y_2)$. If M_T is the matrix that describes the transformation, then

$$M_T \cdot \begin{bmatrix} 1 \\ 0 \end{bmatrix} = \begin{bmatrix} x_1 \\ y_1 \end{bmatrix}$$

TABLE 8.1.
Algebraic and Matrix Representations

Transformation	Mapping Rule	Matrix
Reflection		
In the x-axis	$(x, y) \to (x, -y)$	$\begin{bmatrix} 1 & 0 \\ 0 & -1 \end{bmatrix}$
In the y-axis	$(x, y) \to (-x, y)$	$\begin{bmatrix} -1 & 0 \\ 0 & 1 \end{bmatrix}$
In the line $y = x$	$(x, y) \to (y, x)$	$\begin{bmatrix} 0 & 1 \\ 1 & 0 \end{bmatrix}$
In the line $y = -x$	$(x, y) \to (-y, -x)$	$\begin{bmatrix} 0 & -1 \\ -1 & 0 \end{bmatrix}$
Rotation		
90° about (0, 0)	$(x, y) \to (-y, x)$	$\begin{bmatrix} 0 & -1 \\ 1 & 0 \end{bmatrix}$
180° about (0, 0)	$(x, y) \to (-x, -y)$	$\begin{bmatrix} -1 & 0 \\ 0 & -1 \end{bmatrix}$
270° about (0, 0)	$(x, y) \to (y, -x)$	$\begin{bmatrix} 0 & 1 \\ -1 & 0 \end{bmatrix}$
Translation		
h units horizontally and k units vertically	$(x, y) \to (x + h, y + k)$	No matrix representation
Dilation		
centered at (0, 0) with scale factor k	$(x, y) \to (kx, ky)$	$\begin{bmatrix} k & 0 \\ 0 & k \end{bmatrix}$
Stretch		
centered at (0, 0), a units horizontally and b units vertically	$(x, y) \to (ax, by)$	$\begin{bmatrix} a & 0 \\ 0 & b \end{bmatrix}$

and

$$M_T \cdot \begin{bmatrix} 0 \\ 1 \end{bmatrix} = \begin{bmatrix} x_2 \\ y_2 \end{bmatrix}$$

or, combining,

$$M_T \cdot \begin{bmatrix} 1 & 0 \\ 0 & 1 \end{bmatrix} = \begin{bmatrix} x_1 & x_2 \\ y_1 & y_2 \end{bmatrix}.$$

Observing, however, that

$$\begin{bmatrix} 1 & 0 \\ 0 & 1 \end{bmatrix}$$

is the identity matrix tells us what M_T must be, namely,

$$M_T = \begin{bmatrix} x_1 & x_2 \\ y_1 & y_2 \end{bmatrix}.$$

Stated generally, this result says that the columns of the matrix for a linear transformation are the coordinates of the images of the points (1, 0) and (0, 1) under that transformation.

As an application of the result above, consider the rotation of 90 degrees counterclockwise (+90°) about the origin. Using the mapping rule $(x, y) \rightarrow (-y, x)$, or a simple sketch such as that in figure 8.2, we see that $(1, 0) \rightarrow (0, 1)$ and that $(0, 1) \rightarrow (-1, 0)$. Thus, the matrix that describes a rotation of 90° about the origin is

$$M_{90} = \begin{bmatrix} 0 & -1 \\ 1 & 0 \end{bmatrix}.$$

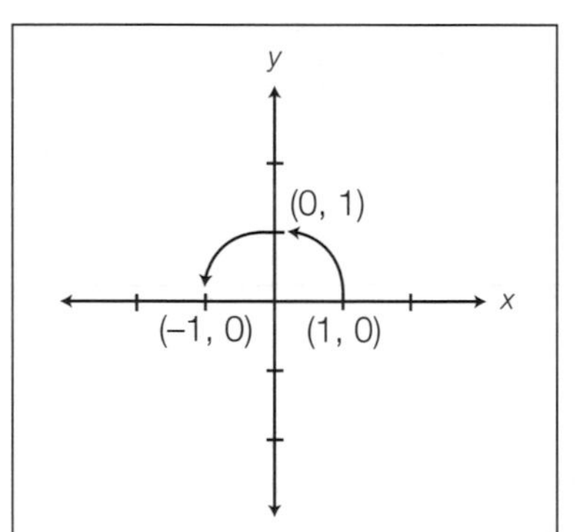

Fig. 8.2. Rotation of 90° counterclockwise about the origin

The ease with which the matrices representing the linear transformations can be derived means that students do not have to memorize long lists of matrix descriptions. They can easily reconstruct any matrix merely by knowing the images of the two points (1, 0) and (0, 1)!

COMPOSITIONS

Interesting activities that explore the composition of transformations can be developed using their matrix representations. What is the net effect, for example, when a reflection in the x-axis is followed by a rotation of 90 degrees? Students should conjecture what they think will happen, then use the matrices to determine the validity of their conjecture. Using the matrices in table 8.1, we would represent this example algebraically as

$$\begin{bmatrix} 0 & -1 \\ 1 & 0 \end{bmatrix} \cdot \begin{bmatrix} 1 & 0 \\ 0 & -1 \end{bmatrix} \cdot \begin{bmatrix} x \\ y \end{bmatrix} = \begin{bmatrix} ? \\ ? \end{bmatrix}.$$

The resulting ordered pair (y, x) tells us that this composition is the same as a reflection in the line $y = x$.

A particularly intriguing composition result is that *every rotation, translation, and glide reflection is equivalent to a composition of reflections.* Although this *general* case cannot be established using matrices—we can represent only *specific* isometries using two-by-two matrices—the result is easily determined using the properties of reflections. For example, the composition of two reflections, with reflection lines intersecting at an angle of $\theta/2$ degrees, is the same as a rotation of θ degrees around the point of intersection. The angle of rotation is twice the measure of the angle formed by the intersection of the two lines. The direction of the rotation is determined by the line that is used for the first reflection. This is depicted in figure 8.3 where P maps to P'' through a reflection in L_1 followed by a reflection in L_2.

Using properties of reflections and some right-triangle geometry, we can show that $AP = AP''$ and that the angle PAP'' is θ degrees for any point P. Hence P'' is the image of P through a rotation of θ degrees about A. Similarly, a translation can be described by the composition of two reflections through two parallel lines. (The exact relationship between the two lines of reflection and the direction and distance of the translation is left to the reader and to the reader's students.) Furthermore, using the composite information about the translation, it is easily shown that the glide reflection, itself a composition of a translation and a reflection, is equivalent to the composition of *three* reflections.

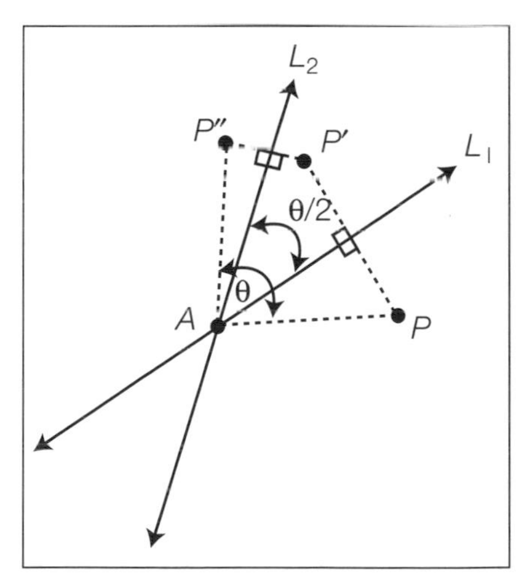

Fig. 8.3. The composition of two reflections in lines intersecting at a point

CONIC SECTIONS

Most high school students are familiar with the quadratic equation $y = x^2$ and its graph. When faced, however, with equations of the form $y = ax^2 + bx + c$, $a \neq 0$, or more challengingly, $x = ay^2 + by + c$, students are often at a loss to interpret the information such equations can provide. One way to overcome this hurdle is to use a transformational approach on the quadratic function and on its geometric counterpart, the parabola. By writing the equation of the parabola in transformational form, $(y - k) = \pm a(x - h)^2$, students are able to relate each new equation to the initial building block, $y = x^2$. Connections are formed. Students come to realize that there is really only one basic shape!

The graphing calculator is an excellent tool for students to use to explore the meaning of the constants in the transformational form given above. (This can also be done by hand, but it takes more time.) For example, students can be asked to plot on the same axes $y = x^2$, $y = (x - 3)^2$, $y = (x - 5)^2$, $y = (x + 4)^2$, $y = (x + 0.5)^2$, and so on. After graphing a series of such examples, they should be asked to make conjectures about the effect of the added term on the graph of $y = x^2$. (It shifts the graph in a horizontal direction.) Similar activities can be used to explore the effect of adjusting the y-value (e.g., comparing the graph of $y = x^2$ with examples such as $y - 4 = x^2$, $y - 1.5 = x^2$, $y + 3 = x^2$, $y + 6 = x^2$). Finally, students should combine their results to establish that the graph of the equation $y - k = (x - h)^2$ is merely the parabola $y = x^2$ translated h units in the horizontal direction and k units in the vertical direction. Thus, for example, rewriting $y + 3 = (x - 4)^2$ as $y - (-3) = (x - 4)^2$ indicates that the graph of this function is the graph of $y = x^2$ translated *down* 3 units and *right* 4 units. (See the article by Hirschhorn and Viktora for further discussion of transformations with graphing calculators.)

A similar approach can help students discover the nature of the graph of $y = ax^2$ when values of a other than 1 are selected, that is, a stretch of factor a in the vertical direction. While developing this, students should be asked to explore a range of related ideas: Does the graph open up or down? How does the direction of opening relate to the coefficient of x^2? Is the vertex the highest point or the lowest point on the graph? Note that when a is negative, it is usually easier to interpret its effect as a stretch of $|a|$ followed by a reflection in the x-axis, or vice versa. (The interested reader may wish to prove, alternatively, that the stretch of $|a|$ is equivalent to a stretch in the *horizontal* direction of

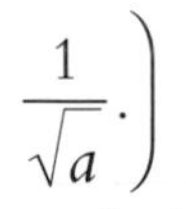

Combining the findings above results in the transformational form of the quadratic equation (see fig. 8.4). As the following example shows, the quadratic equation written in this form provides the student with immediately accessible and interpretable information.

Example: For the parabola

$$y = -\tfrac{1}{2}x^2 + 3x + 1,$$

a. give the coordinates of the vertex;
b. indicate whether the parabola opens up or down;
c. sketch the graph.

Solution:

Expressing the function in transformational form, we find that

$$y = -\frac{1}{2}x^2 + 3x + 1 \text{ becomes } \left(y - \frac{11}{2}\right) = -\frac{1}{2}(x-3)^2.$$

Thus,

a. the vertex is $\left(3, \frac{11}{2}\right)$;
b. the parabola opens downward;
c. $y = -\frac{1}{2}x^2 + 3x + 1$ is $y = x^2$ under—

- a reflection in the x-axis;
- a vertical stretch of $\frac{1}{2}$;
- 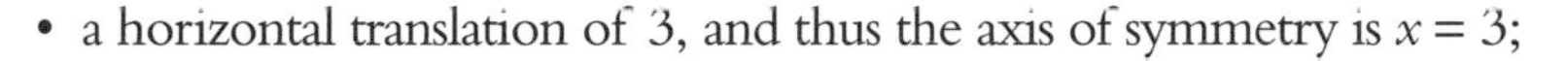
a horizontal translation of 3, and thus the axis of symmetry is $x = 3$;
- a vertical translation of $\frac{11}{2}$.

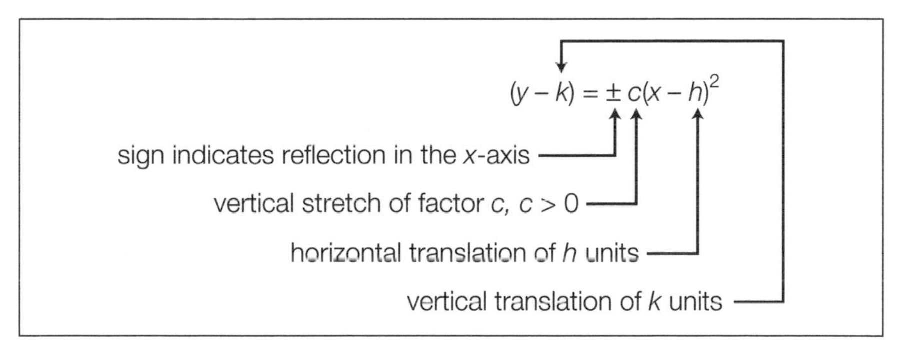

Fig. 8.4

Two other forms of the parabola that are commonly studied are $x = y^2$ and $-x = y^2$, the parabolas that open right and left, respectively. These can be derived by rotating $y = x^2$ through an angle of ±90° when the turn center is (0, 0). To illustrate, recall that a rotation of 90 degrees maps $(x, y) \rightarrow (-y, x)$. Thus, under this rotation, $y = x^2$ maps to $x = (-y)^2 = y^2$. In a manner similar to that above, or by using properties of rotations, students can explore the interpretation of the general transformational form for these equations: $(x - h) = \pm\, c(y - k)^2$.

The more difficult case of parabolas that do not have an axis of symmetry parallel to either the x- or y-axis will not be developed, but the equations of these, too, can be derived and interpreted using transformations. (For a discussion of this general case, see Wooton, Beckenbach, and Fleming 1972.) The second-degree equation, $ax^2 + 2hxy + by^2 + 2gx + 2fy + c = 0$, like all the conic sections, is best approached by first considering which rotation about the origin will eliminate the xy term. The resulting equation represents a graph with an axis of symmetry parallel to an axis. It can, therefore, be written in standard transformational form and studied further for its properties.

Transformations are also useful when studying the relationships between figures. For example, the two-way stretch, represented by the mapping rule $(x, y) \rightarrow (ax, by)$ illuminates several relationships between the circle and the ellipse. The most obvious of these, shown below, is that the circle and the ellipse are transformations of each other.

Consider the unit circle, $x^2 + y^2 = 1$, under the scale transformation $(x, y) \rightarrow (ax, by)$, $a \neq 0$, $b \neq 0$. Let $x' = ax$ and $y' = by$. Then $x = x'/a$ and $y = y'/b$. When these values are substituted into the equation for the circle, the result is $(x'/a)^2 + (y'/b)^2 = 1$. Students can study the shape of this new figure by trying different values for a and b in the equation and considering what these values mean in terms of the stretches that they generate on the circle. (The scale factors also categorize the relationship between the area of the unit circle, π, and the area of the resulting ellipse, $ab\pi$. Limited space, however, does not allow for that development here.)

TRIGONOMETRY

Students' first introduction to trigonometry is usually as an extension of the Pythagorean theorem. Trigonometric ratios are defined using the sides and the hypotenuse of a right triangle. Equivalent trigonometric ideas can, of course, also be developed on the Cartesian plane where the image of the point (1, 0) after a rotation about the origin of θ degrees counterclockwise is defined to be the point $(\cos \theta, \sin \theta)$ (see fig. 8.5). Although it will not be proved here, the image of the point (0, 1) under a rotation of $\theta°$ is $(-\sin \theta, \cos \theta)$. From the relationship established in the matrix section, we know that these two mappings determine the general matrix for a rotation of θ degrees about the origin, namely,

$$M_\theta = \begin{bmatrix} \cos\theta & -\sin\theta \\ \sin\theta & \cos\theta \end{bmatrix}.$$

In turn, this general rotation matrix can be used to find the image of any point (x, y) under a rotation of any degree measure:

$$\begin{bmatrix} \cos\theta & -\sin\theta \\ \sin\theta & \cos\theta \end{bmatrix} \cdot \begin{bmatrix} x \\ y \end{bmatrix} = \begin{bmatrix} a \\ b \end{bmatrix}$$

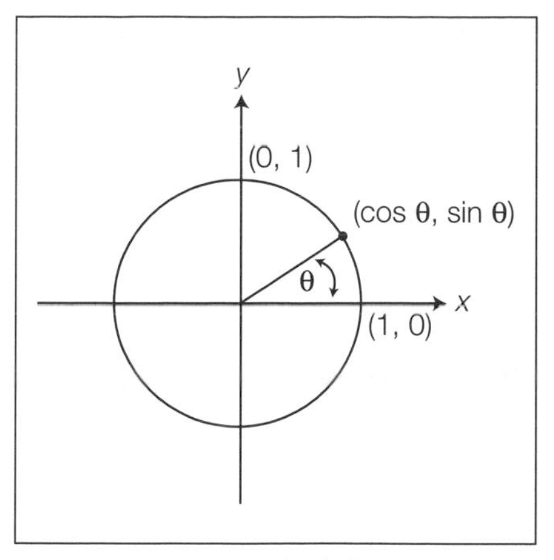

Fig. 8.5. Unit circle definition of trigonometric functions

One of the more elegant applications of the general rotation above establishes the trigonometric identities for cos (α + β) and sin (α + β). This is accomplished, as shown below, by determining the image of the point (1, 0) after a rotation of α + β degrees, call it $T_{(\alpha+\beta)}$ (1, 0), in two different ways. The resulting statements are then compared.

First consider a single rotation of (α + β) degrees (fig. 8.6a). In this case, using the matrix above with θ = (α + β), we get

$$T_{(\alpha+\beta)}(1, 0) = \begin{bmatrix} \cos(\alpha+\beta) & -\sin(\alpha+\beta) \\ \sin(\alpha+\beta) & \cos(\alpha+\beta) \end{bmatrix} \cdot \begin{bmatrix} 1 \\ 0 \end{bmatrix} = \begin{bmatrix} \cos(\alpha+\beta) \\ \sin(\alpha+\beta) \end{bmatrix}. \quad (1)$$

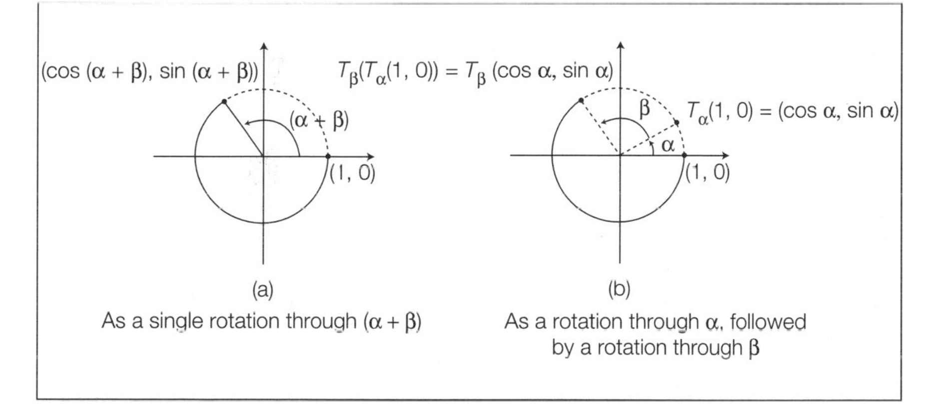

Fig. 8.6. Rotation through two angles, α and β

Next consider the image of (1, 0) after a rotation of α degrees:

$$T_{\alpha}(1, 0) = (\cos \alpha, \sin \alpha)$$

Then consider the image of that point after a rotation of β degrees, that is, $T_{\beta}(T_{\alpha}(1, 0))$ (fig. 8.6b):

$$T_{\beta}\left(T_{\alpha}(1,0)\right) = \begin{bmatrix} \cos\beta & -\sin\beta \\ \sin\beta & \cos\beta \end{bmatrix} \cdot \begin{bmatrix} \cos\alpha \\ \sin\alpha \end{bmatrix} = \begin{bmatrix} \cos\beta\cos\alpha - \sin\beta\sin\alpha \\ \sin\beta\cos\alpha + \sin\alpha\cos\beta \end{bmatrix} \quad (2)$$

The matrices in (1) and (2) represent the same point. Each is an image of the point (1, 0) rotated through the same number of degrees, although in the first case the rotation was accomplished in a single rotation of $(\alpha + \beta)$ degrees and in the second case the rotation was completed in two steps, a rotation of α degrees followed by a rotation of β degrees. Thus the matrices are equal:

$$\begin{bmatrix} \cos(\alpha + \beta) \\ \sin(\alpha + \beta) \end{bmatrix} = \begin{bmatrix} \cos\beta\cos\alpha - \sin\beta\sin\alpha \\ \sin\beta\cos\alpha + \sin\alpha\cos\beta \end{bmatrix}$$

And, from equality of matrices, we have the desired addition formulas:

$$\cos(\alpha + \beta) = \cos\beta\cos\alpha - \sin\beta\sin\alpha$$
$$\sin(\alpha + \beta) = \sin\beta\cos\alpha + \sin\alpha\cos\beta$$

CONCLUSIONS

Advocates of the inclusion of transformations in the school curriculum state that transformations supply a unifying concept not only in the study of Euclidean geometry but among many other branches of mathematics as well. Some examples of this connectivity have been presented above. Additional areas where transformations could be used include inverse functions; symmetries in graphs (e.g., translation symmetry); vectors as translations; polar coordinates introduced through dilations and rotations; the translation and rotation of axes, particularly as they affect the conic sections; and the study of logarithms.

Using transformations to study mathematics has many advantages. It exemplifies the interconnectedness of apparently distinct mathematical areas. It reinforces the beauty and logic of mathematics as a discipline. It demonstrates the utility and versatility of mathematical thinking. In the long run, it may even make the teacher's job easier, since students will have acquired prerequisite knowledge that expedites the study of "new" topics. But, most important, this approach promotes student learning.

REFERENCES

National Council of Teachers of Mathematics. *Curriculum and Evaluation Standards for School Mathematics.* Reston, Va.: The Council, 1989.

Wooton, William, Edwin F. Beckenbach, and Frank J. Fleming. *Modern Analytic Geometry.* Boston, Mass.: Houghton Mifflin Co., 1972.

FOR FURTHER READING

Coxford, Arthur F., Jr. *Geometry from Multiple Perspectives. Curriculum and Evaluation Standards for School Mathematics* Addenda Series, Grades 9–12, edited by Christian R. Hirsch. Reston, Va.: National Council of Teachers of Mathematics, 1991.

Okolica, Steve, and Georgette Macrina. "Integrating Transformation Geometry into Traditional High School Geometry." *Mathematics Teacher* 85 (December 1992): 716–19.

Rising, Gerald R., John A. Graham, John G. Balzano, Janet M. Burt, and Alice M. King. *Unified Mathematics, Book 3.* Boston, Mass.: Houghton Mifflin Co., 1989.

9

Using Transformations to Foster Connections

Daniel B. Hirschhorn
Steven S. Viktora

WHAT makes a unit of school mathematics desirable? An answer to this question might include such qualities as the content's being widely applicable, rich in connections to other mathematics, intuitive at the outset, elegant in presentation, consistent with the *Curriculum and Evaluation Standards for School Mathematics* (National Council of Teachers of Mathematics [NCTM] 1989), and, perhaps, easy or interesting to learn and teach. Transformations fit such a list. As shown in the preceding two chapters of this yearbook, the study of transformations begins with notions that all students know. Transformations connect algebra and geometry in many, often elegant ways, and they are widely applicable. In this article, we focus on the qualities of transformations that can be used in secondary school mathematics classes.

TRANSFORMATIONS IN ALGEBRA AND GEOMETRY

Transformations are geometric functions, and as stated in the *Standards,* "the concept of function is an important unifying idea in mathematics" (NCTM 1989, p. 154). Transformations give students a visual introduction to functions and provide a solid foundation for later mathematical ideas. Simple mappings on the coordinate plane are understandable and connect with more sophisticated mathematics. Reflecting a point over the y-axis leads to the generalization $(x, y) \rightarrow (-x, y)$. That generalization will later lead to the idea of even functions. A rotation of 180° about the origin is generalized as $(x, y) \rightarrow (-x, -y)$ and leads to the idea of odd functions. The mapping $(x, y) \rightarrow (y, x)$ describes a reflection over the line $y = x$ and serves as an elegant way to introduce inverses of functions.

In addition to the familiar distance-preserving transformations, or *isometries, similarity transformations* offer an intuitive approach to similarity. For instance, the simple mapping $(x, y) \to (kx, ky)$ describes a size change centered at the origin. The composition of isometries and similarity transformations leads to a more general view of congruence and similarity: *Two figures are similar if and only if one is the image of the other under a reflection, rotation, translation, size change, or any composite of these. Two figures are congruent if and only if one is the image of the other under a reflection, rotation, translation, or any composite of these.* Thus, congruence is a special case of similarity. Most important, congruence and similarity are not restricted to triangles; they apply to all figures in all dimensions!

On a coordinate plane, mapping all points (x, y) of a figure to (kx, ky) results in a similar figure. The mapping is a size change whose center of dilation is the origin and whose size-change factor is $|k|$. Similarity transformations also can provide real applications and further connections to other mathematics, such as in producing scale drawings, model constructions, and fractal images. (For a more thorough treatment of the subject, see Senk and Hirschhorn [1990].) Single-direction stretches can visually be compared and contrasted to size changes. For example, mapping all points (x, y) of a figure to (x, ky) is a vertical stretch whereas mapping (x, y) to (kx, y) is a horizontal stretch.

TRANSFORMATIONS IN UPPER-LEVEL COURSES

Many wonderful results can be proved elegantly using transformations. For instance, the size change determined by

$$(x, y) \to \left(\frac{1}{a}x, \ \frac{1}{a}y\right)$$

can also be described symbolically as

$$x_{\text{new}} = \frac{1}{a} \cdot x_{\text{old}}$$

and

$$y_{\text{new}} = \frac{1}{a} \cdot y_{\text{old}}.$$

Solving for the original variables yields

$$x_{\text{old}} = a \cdot x_{\text{new}}$$

and

$$y_{\text{old}} = a \cdot y_{\text{new}}.$$

If these replacements are used in the equation $y = x^2$, the result is $ay = (ax)^2$. Thus, by the definition of similarity, the graph of $y = x^2$ is similar to that of $y = ax^2$. As shown in chapter 8, mapping $(x, y) \to (x + h, y + k)$

produces a translation of the graph of the parabola. Thus, it is established that all parabolas are similar!

Suppose that the graph of $y = x^2$ is translated by the mapping $(x, y) \rightarrow (x + h, y + k)$. Students easily comprehend that for each point on the graph, h is added to the x-coordinate and k is added to the y-coordinate. But it is difficult for students to comprehend that in the equation for the new graph, $(y - k) = (x - h)^2$, the parameters are subtracted. The following discussion outlines a successful lesson with this objective.

In order for students to learn how translations affect graphs and their equations, they must generate and graph many cases. Technology is a great aid in helping students analyze and create graphs. With graphing calculators or a computer graphing program, students can plot a variety of equations with different parameters. This lesson uses the program TRNSGRPH for the Texas Instruments TI-82 graphing calculator. (The appendix to this article lists programming code for this program as well as others that will reflect, translate, or rotate either the graph of a function or a polygon.) With the capabilities of technology, programs can be electronically transferred onto the students' own calculators; teachers can do so the day before a lesson or in just a few minutes while students are correcting homework or working on an opening activity.

The lesson starts with the question of transforming the graph of $y = x^2$ by the mapping $(x, y) \rightarrow (x + 5, y - 3)$. Students (in pairs or in groups) are charged with performing this mapping and answering the following questions, which are projected on a screen or written on the chalkboard:

a) What is the shape of the image?

b) Is the image congruent to the preimage? Why or why not?

c) Guess an equation that gives the graph of the image.

This part of the lesson sets up the use of the programs by reviewing the point-by-point mapping of translations. After the students have worked on these questions, they discuss the answers to the first two questions and conclude that the mapping describes a translation, which is an isometry, so the image is a parabola congruent to the preimage.

Next the students check their guesses by using the program TRNSGRPH. In using this program, the students should adjust their calculators (using the ZSTANDARD command followed by the ZSQUARE command under the ZOOM menu on a TI-82) so that the graphs do not appear distorted. In running the program, students are first asked for the preimage equation as Y_1. Then students must put in the translation vector (h, k). In this case, $h = 5$ and $k = -3$. The program graphs the preimage first and then the image. Then the students type their guesses for the equation as Y_2 (see fig. 9.1). The program graphs the equations of the guess, the preimage, and the image.

The student has made the common guess that the equation of the image is $(y - 3) = (x + 5)^2$. In figure 9.1, 3 is added to both sides to

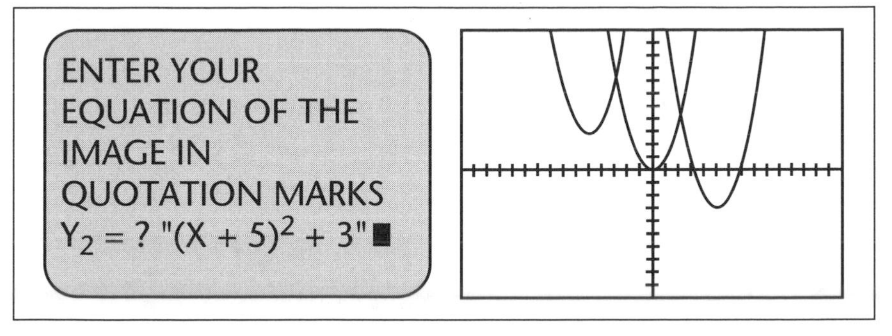

Fig. 9.1

produce $Y_2 = (x + 5)^2 + 3$. The three distinct graphs on the right-hand screen demonstrate that the guess is not correct. For a correct guess, only the preimage and image graphs appear. Sometimes the teacher's help is needed to give students ideas for refining wrong guesses, but this happens more rarely than one might expect.

When students complete this first case, they should do several more examples using different translation vectors and different preimages. They should conclude the activity by generalizing their results, and many students have demonstrated that they are able to do so. Further, having done the hands-on activities using technology, students are able to appreciate an algebraic derivation of the effect translations have on the equations of graphs.

A wonderful connection to, and payoff of, learning translations of graphs is in its applications to statistics and interpreting measures of central tendency (Rubenstein et al. 1992). One translation of a set of data $\{x_1, x_2, ..., x_n\}$ is a transformation that maps each x_i to $x_i + h$, where h is a constant. Using transformation (function) notation yields $T(x) = x + h$. In the following example, students can work through how a translation of data affects the range, mean, median, mode, and standard deviation of a data set:

> Suppose that a small company employs 9 part-time student employees. The normal numbers of hours worked by these employees in a week are 18, 20, 22, 15, 20, 20, 15, 22, and 14. Because of a special sale and a school holiday, each employee is asked to work 10 extra hours. Thus the numbers of hours worked by the 9 employees in this week will be 28, 30, 32, 25, 30, 30, 25, 32, and 24.

Graphing the frequency distribution of the hours in the different weeks gives students a visual picture of the translation (see fig. 9.2).

By computing the mean, median, and mode of both data sets, students will realize that all three are translated by 10 units as well, whereas the values of the range and standard deviation are the same for the two data sets. The knowledge of translations allows students to observe that the translation of a figure gives a congruent figure. They can then make the simple connection that when a data set is translated, the numbers combined with the pictures show that the center is translated but the spread of

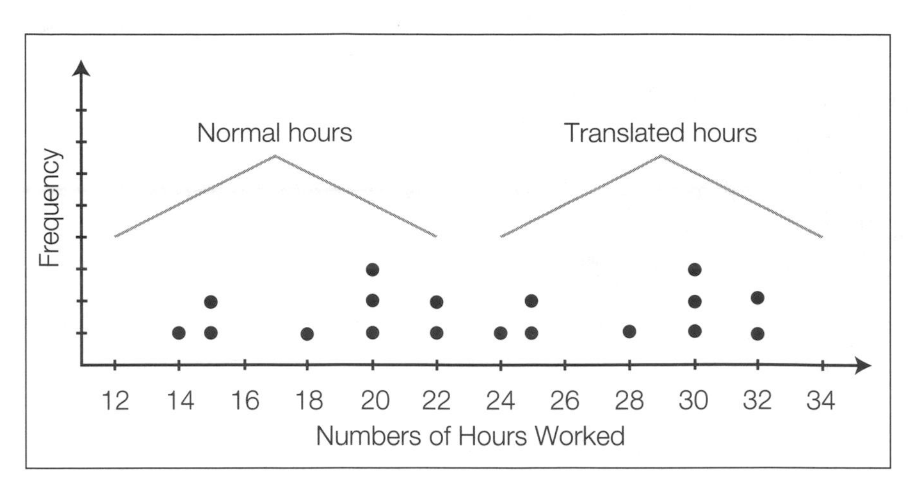

Fig. 9.2

the data remains constant. Students need several examples to help them see the generalization that adding h to each number in a data set adds h to the mean, median, and mode but does not change the range, interquartile range, variance, or standard deviation of the data.

Many real applications of scaling data help students connect transformations and statistics to something meaningful in their lives. A nice example is comparing college-placement tests. Suppose that on Test A the mean is 18.8 and the standard deviation is 5.9 and that on Test B the mean is 500 and the standard deviation is 100. To compare different scores on the different tests, the students can multiply the Test A data by the scale factor 100/5.9, giving the scaled Test A data a mean of 318.64 and a standard deviation of 100. Students can take their own scores for the different tests and compare the distances of their scores from the mean.

In general, multiplying all terms of a data set by a scale factor k multiplies the mean, median, and mode by k; the standard deviation and the range by $|k|$; and the variance by k^2. The example above is also a nice problem-solving activity, since students can solve it without scale changes by seeing how many standard deviations their test scores are from the mean. Discussing several ways to do problems and when transformations are appropriate helps students to make and remember connections. Other good examples of scale changes to data are found in economic indicators, such as the Consumer Price Index (CPI), a measure of inflation that compares prices of a given collection of products at different points in time, and the "dating index," a measure of inflation in the cost of social activities such as dinner and a movie.

Scaling of data also occurs when semi-log or log-log graph paper is used to plot a data set. For instance, to graph a data set of the population of the United States by census year on semi-log paper, the mapping $(x, y) \rightarrow (x, \log y)$ is used. This exercise introduces students to a commonly applied transformation that is not a similarity transformation. Rather, this transformation is used to make exponential curves appear linear. By exploring data sets and graphs on semi-log or log-log paper (or on graphing calculators that have the capability of producing semi-log and log-log graphs), students can connect statistics, functions, powers, and geometry.

Transformations furnish not only connections to real applications but also some of the most elegant connections within the realm of mathematics. A nice connection is in finding areas. Take $\triangle TRI$ defined by $T = (2, 3)$, $R = (0, 0)$, and $I = (4, 0)$. As shown in the leftmost graph of figure 9.3, the area of $\triangle TRI$ is clearly 6. Apply the scale change

$$(x, y) \rightarrow \left(x, \frac{y}{6}\right)$$

to $\triangle TRI$ (see the middle graph of fig. 9.3). The area of the image $\triangle T'R'I'$ is also divided by 6 and thus equals 1. If the scale change

$$(x, y) \rightarrow \left(2x, \frac{y}{6}\right)$$

is applied to $\triangle TRI$ (see the rightmost graph of fig. 9.3), the area of the image $\triangle T''R''I''$ is multiplied by 2 and divided by 6 and thus equals 2. Exploring a number of examples involving different geometric figures can lead to the discovery of the following theorem: *If the scale change $(x, y) \rightarrow (ax, by)$ is applied to a region, the area of the image is ab times the area of the preimage.* One elegant consequence of this theorem is that the area of the ellipse

$$\frac{x^2}{a^2} + \frac{y^2}{b^2} = 1$$

is πab, since the ellipse is a transformation of the unit circle under the scale change $(x, y) \rightarrow (ax, by)$. Another important result is that when the sides of a polygon are transformed by a scale factor k in the mapping $(x, y) \rightarrow (kx, ky)$, the area of the polygon is increased by k^2.

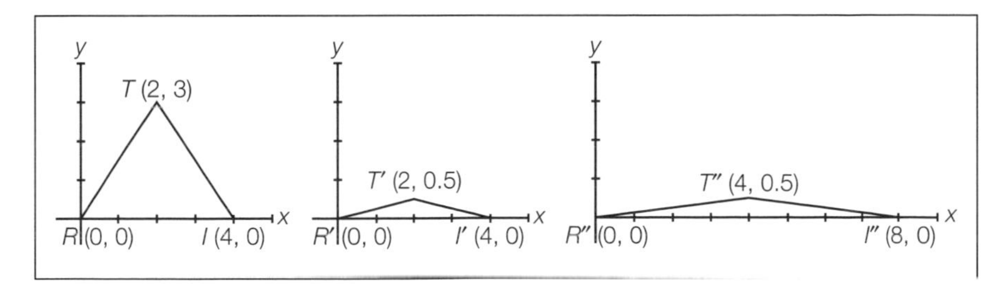

Fig. 9.3

The theorem above also connects transformations to statistics. Consider the function $y = e^{-x^2}$ graphed in figure 9.4. It is proved in calculus that

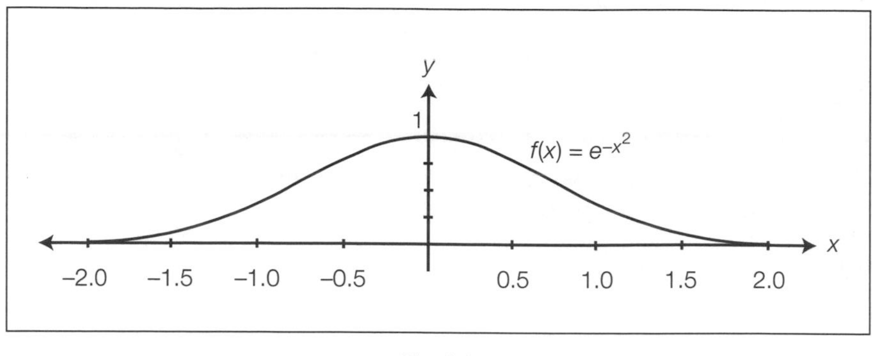

Fig. 9.4.

the area under this curve is $\sqrt{\pi}$ and that the points of inflection are at

$$x = \frac{\pm 1}{\sqrt{2}}.$$

The scale change $(x, y) \rightarrow (x\sqrt{2}, y)$ ensures that the points of inflection are at ±1, but it also changes the area under the curve to $\sqrt{2\pi}$. The equation of the image can be written as

$$y = e^{-\left(\frac{x}{\sqrt{2}}\right)^2} = e^{-\frac{x^2}{2}}.$$

The scale change

$$(x, y) \rightarrow \left(x\sqrt{2}, \frac{y}{\sqrt{2\pi}}\right)$$

will make the area under the curve equal to 1, and the equation of the image will be

$$y\sqrt{2\pi} = e^{-\frac{x^2}{2}}.$$

This last equation gives the standard normal curve!

Transformations also offer an elegant connection to complex numbers. Adding $3 + i$ to $-2 + 4i$ can be considered the translation determined by the vector (3, 1) on the point (−2, 4). Multiplying $-2 + 4i$ by 3 can be considered a size change of magnitude 3 on the point (−2, 4). Multiplying $-2 + 4i$ by $5 - 3i$ can be considered the composition of a size change of magnitude

$$\sqrt{5^2 + (-3)^2}$$

and a rotation of $\tan^{-1}(-3/5)$.

CONCLUSIONS

The purpose of this article is to glimpse the elegant connections that can be made by developing mathematics through transformations. Although the exercises can range from the simple process of flipping a point over a line to the sophisticated derivation of the normal curve, all of them are within the grasp of students. Transformations are a wonderful way to develop and expand on the students' intuition, which leads students to make connections and to develop their understandings of mathematics and its uses.

REFERENCES

National Council of Teachers of Mathematics. *Curriculum and Evaluation Standards for School Mathematics.* Reston, Va.: The Council, 1989.

Rubenstein, Rheta N., et al. *Functions, Statistics, and Trigonometry.* Glenview, Ill.: ScottForesman, 1992.

Senk, Sharon L., and Daniel B. Hirschhorn. "Implementing the *Standards:* Multiple Approaches to Geometry: Teaching Similarity." *Mathematics Teacher* 84 (April 1990): 274–80.

APPENDIX

TRANSFORMATION PROGRAMS FOR THE TEXAS INSTRUMENTS TI-82 GRAPHING CALCULATOR

The first set of programs performs the transformations of reflection, translation, and rotation on graphs. The second set of programs performs the same transformations on polygons.

Before running any of the programs, the viewing window should be set so that $(X_{max} - X_{min}) = 1.5\ (Y_{max} - Y_{min})$. Doing so "squares off" the display and avoids the appearance of a lack of congruence between the preimage and the image. (The command ZSQUARE under the ZOOM menu is useful for squaring off.) The programs will run with either degrees or radians, but it is necessary to enter the desired mode before running a program.

The first three programs, ENTERY1, ENTERY2, and PLOTGRPH are all subroutines called by the main transformation programs.

PROGRAM: ENTERY1

```
:ClrHome
:ClrDraw
:FnOff
:Disp "ENTER Y1 IN"
:Disp "QUOTATION MARKS"
:Prompt Y1
```

PROGRAM: ENTERY2

```
:ClrHome
:Disp "ENTER YOUR"
:Disp "EQUATION OF THE"
:Disp "IMAGE IN"
:Disp "QUOTATION MARKS"
:Prompt Y2
```

PROGRAM: PLOTGRPH

```
:ClrDraw
:DispGraph
:1→I
:500→N
:(Xmax–Xmin)/(N–1)→S
:Xmin→X
:Lbl 1
:X→[A](1,1)
:Y1→[A](2,1)
:[B][A]→[C]
:Pt-On([C](1,1), [C](2,1))
:X+S→X
:IS>(I,N)
:Goto 1
```

PROGRAM: REFGRPH reflects the graph of any function over any line through the origin and also graphs the mirror line.

PROGRAM: REFGRPH

Line	
1	:ClrDraw
2	:prgmENTERY1
3	:Disp "REFLECT OVER"
4	:Disp "Y=MX or X=0"
5	:Disp "If Y=MX, PRESS 1"
6	:Disp "If X=0, PRESS 2"
7	:Input A
8	:If A=1
9	:Then
10	:Prompt M
11	:tan^{-1} M→θ
12	:{2,1}→dim [A]
13	:{2,2}→dim [B]
14	:{2,1}→dim [C]
15	:cos (2θ)→[B](1,1)
16	:sin (2θ)→[B](1,2)
17	:sin (2θ)→[B](2,1)
18	:–cos (2θ)→[B](2,2)
19	:[B][A]→[C]
20	:prgmPLOTGRPH
21	:Pause
22	:DrawF MX
23	:Stop
24	:End
25	:If A=2
26	:Then
27	:{2,2}→dim [B]
28	:–1→[B](1,1)
29	:0→[B](1,2)
30	:0→[B](2,1)
31	:1→[B](2,2)
32	:[B][A]→[C]
33	prgmPLOTGRPH

Line 2. This instruction allows the user to enter the equation of the function.
Lines 3, 4, 5, 6. These lines prompt the user to tell whether the mirror line has a slope or if the mirror line is the *y*-axis.
Line 7. The user enters "1" or "2."

Line 10. If the user chose "1" at line 7, the program asks for the slope of the mirror line. After the user presses ENTER, the calculator will graph the preimage and the image.

Lines 21, 22. At this "Pause" statement, the user should press ENTER to make the calculator graph the mirror line.

Line 25. If the user chose option "2" at line 7, the program comes here to graph the reflection over the *y*-axis.

PROGRAM: ROTGRPH rotates the graph of any function about the origin.

PROGRAM: ROTGRPH

Line	
1	:ClrDraw
2	:prgmENTERY1
3	:Disp "MAGNITUDE OF"
4	:Disp "ROTATION"
5	:Input θ
6	:{2,2}→dim [B]
7	:{2,1}→dim [A]

Line 2. This instruction allows the user to enter the equation of the function.
Lines 3, 4, 5. These instructions allow the user to enter the magnitude of the rotation.
However, the program will use whatever angle mode the user had entered into the calculator before calling the program.

```
8    :{2,1}→dim [C]
9    :cos θ→[B](1,1)
10   :sin θ→[B](2,1)
11   :-sin θ→ [B](1,2)
12   :cos θ→ [B](2,2)
13   :prgmPLOTGRPH
```

PROGRAM: TRNSGRPH translates the graph of any function. After drawing the graphs of the preimage and the image, it asks the user to enter a proposed equation for the image. The program then graphs this as well, enabling the user to see if the equation is correct. (This guess feature can also be added to REFGRPH and ROTGRPH by inserting the last four lines below into those programs.)

PROGRAM: TRNSGRPH

Line		
1	:prgmENTERY1	Line 1. This instruction allows the user to enter the equation of the function.
2	:{2,2}→ dim [B]	
3	:{2,1}→ dim [A]	
4	:{2,1}→ dim [C]	
5	:[A]→ [C]	
6	:Disp "TRANSLATION OF"	Lines 6, 7, 8, 9. These lines prompt the user to state that the preimage is to be translated *H* units horizontally and *K* units vertically. After pressing ENTER, the user will will get the graphs of the preimage and the image.
7	:Disp "(H,K)"	
8	:Prompt H	
9	:Prompt K	
10	:{3,1}→ dim [A]	
11	:{3,1}→ dim [C]	
12	:1→ [A](3,1)	
13	:1→ [C](3,1)	
14	:{3,3}→ dim [B]	
15	:1→ [B](1,1)	
16	:0→ [B](1,2)	
17	:H→ [B](1,3)	
18	:0→ [B](2,1)	
19	:1→ [B](2,2)	
20	:K→ [B](2,3)	
21	:0→ [B](3,1)	
22	:0→ [B](3,2)	Line 28. This instruction allows the user to view the preimage and the image.
23	:1→ [B](3,3)	
24	:[B][A]→ [C]	Line 29. If the user presses ENTER, the program then prompts the user to enter an equation for what she or he thinks is an equation for the image. After pressing ENTER again, the user will see the graphs of the preimage, the image, and his or her guess for the image.
25	:{2,2}→ dim [C]	
26	:prgmPLOTGRPH	
27	:StorePic Pic1	
28	:Pause	
29	:pgrmENTERY2	
30	:RecallPic Pic1	

The programs PLTPOLYG and POLYGENT are subroutines called by the main transformation programs.

PROGRAM: PLTPOLYG

```
:1→I
:Lbl 1
:I+1→L
:Line (([C](1,I)), ([C](2,I)),
  ([C](1,L)), ([C](2,L)))
:IS > (I,N–1)
:Goto 1
:Line (([C](1,L)), ([C](2,L)),
  ([C](1,1)), ([C](2,1)))
```

PROGRAM: POLYGENT

```
:Disp "NUMBER OF SIDES"
:Input N
:{2,N}→dim [A]
:{2,N}→dim [C]
:1→I
:Disp "ENTER POLYGON"
:Lbl 1
:Prompt X
:X→[A](1,I)
:Prompt Y
:Y→[A](2,I)
:IS > (I,N–1)
:Goto 1
:Prompt X
:X→[A](1,N)
:Prompt Y
:Y→[A](2,N)
:[A]→[C]
```

PROGRAM: REFPOLYG reflects polygons over any line through the origin. See PROGRAM: REFGRPH for any needed documentation.

PROGRAM: REFPOLYG

```
:ClrHome
:ClrDraw
:FnOff
:prgmPOLYGENT
:prgmPLTPOLYG
:Pause
:ClrHome
:Disp "REFLECT OVER"
:Disp "Y=MX or X=0"
:Disp "If Y=MX, PRESS 1"
:Disp "If X=0, PRESS 2"
:Input A
:If A=1
:Then
:Disp "M ="
:Input M
:tan⁻¹ M→θ
:{2,2}→dim [B]
:cos (2θ)→[B](1,1)
:sin (2θ)→[B](1,2)
:sin (2θ)→[B](2,1)
:–cos (2θ)→[B](2,2)
:[B][A]→[C]
:prgmPLTPOLYG
:Pause
:DrawF MX
:Stop
:End
:If A=2
:Then
:{2,2}→dim [B]
:–1→[B](1,1)
:0→[B](1,2)
:0→[B](2,1)
:1→[B](2,2)
:[B][A]→[C]
:prgmPLTPOLYG
```

PROGRAM: ROTPOLYG rotates any polygon about the origin. It has the feature of rotating the polygon repeatedly through the angle specified every time the user presses ENTER. To end the program, the user needs to press ON.

PROGRAM: ROTPOLYG

```
:ClrHome
:ClrDraw
:FnOff
:prgmPOLYGENT
:prgmPLTPOLYG
:Pause
:ClrHome
:Disp "MAGNITUDE OF"
:Disp "ROTATION"
:Input θ
:{2,2}→dim [B]
:cos θ→[B](1,1)
:sin θ→[B](2,1)
:–sin θ→[B](1,2)
:cos θ→[B](2,2)
:Lbl 2
:[B][A]→[C]
:prgmPLTPOLYG
:[C]→[A]
:Pause
:Goto 2
:Pause
:End
```

PROGRAM: TRNSPOLG translates any polygon *H* units horizontally and *K* units vertically.

PROGRAM: TRNSPOLG

```
:ClrDraw
:FnOff
:prgmPOLYGENT
:prgmPLTPOLYG
:Pause
:{3,N}→dim [A]
:{3,N}→dim [C]
:1→Z
:Lbl 2
:1→[A](3,Z)
:IS>(Z,N)
:Goto 2
:[A]→[C]
:{3,3}→dim [B]
:ClrHome
:Disp "TRANSLATION OF"
:Disp "(H,K)"
:Prompt H
:Prompt K
:1→[B](1,1)
:0→[B](1,2)
:H→[B](1,3)
:0→[B](2,1)
:1→[B](2,2)
:K→[B](2,3)
:0→[B](3,1)
:0→[B](3,2)
:1→[B](3,3)
:[B][A]→[C]
:{2,N}→dim [C]
:prgmPLTPOLYG
```

10

Connecting Mathematics with Its History: A Powerful, Practical Linkage

Luetta Reimer
Wilbert Reimer

IMAGINE studying music without learning about Beethoven or Mozart. Would anyone teach *Huckleberry Finn* or *Hamlet* without identifying Twain or Shakespeare? Could U.S. government be taught without the story of the Continental Congress? It is just as vital to trace the sources of mathematical principles and name their originators when teaching mathematics. The history of mathematics, including its principles, procedures, and personalities, is often one of the most neglected areas in our teaching of mathematics. Filled with fascinating material, the history of mathematics is a rich resource available to every teacher.

The purpose of this article is twofold: to furnish a rationale for connecting mathematics with its history, and to offer a variety of practical suggestions on how to implement this connection.

Why Should Mathematics Be Taught in Historical Perspective?

History Motivates

Amy rested her chin in her hands, but her eyes were wide open as she listened to Mrs. Cruz. Jose stopped drumming on the table with his pencil and Natasha quit twisting her hair. The roomful of children sat spellbound, captivated by their teacher's voice. It was mathematics period, and Mrs. Cruz was telling a story about Archimedes. When she got to the part where Archimedes was speared to death by a Roman soldier, the students winced and groaned.

The death of Archimedes

"Archimedes was so busy at work on a mathematics problem that he hardly noticed the soldier approaching," Mrs. Cruz observed. "What kind of problem do you think Archimedes was working on?"

After hearing of Archimedes, Amy, Jose, and Natasha were more open to an introduction to geometry. They realized that mathematics is something that real people do—and sometimes even die for.

A Human Endeavor

History integrated into the mathematics classroom reminds students that mathematics is essentially a human endeavor. In this increasingly technological world, students may begin to infer that mathematics is done only by calculators and computers; they may assume that they have no personal need to understand mathematics or perform computational tasks. Modern advances *have* simplified many processes, but problem solving is still essentially a human task. Discoveries have been made because living people had need of them, and human problems are still the catalysts for mathematical experimentation today.

The Power of the Story

If mathematics is something people do, stories of mathematicians doing it can inspire others to do the same. Nothing communicates like a good story; even the most restless or seemingly uninterested student may listen to a story.

Many of the figures in mathematics history lived fascinating lives. There was Sophie Germain, who had to sneak candles into her room so she could study at night against her parents' wishes. There was Galois, shot and killed in a duel at the age of 21, but not before he left an important legacy in group theory. And there was Newton, who started making good grades after a showdown with the school bully. Those were normal people for the most part, although some had extraordinary powers of concentration. Like everyone, they had obstacles to overcome. Their lives are outstanding testimonies to the power of hard work and determination.

Where Did *That* Come From?

For as long as mathematics has been taught, students have reacted to much of the explanation with a puzzled "Why?" In many situations, mathematics history answers this question. History in the mathematics classroom counteracts students' natural reluctance to accept something simply because "that's the way it is." Understanding the origins of certain ideas, such as algebraic notation or "imaginary" numbers, makes students more receptive to even difficult or abstract mathematical procedures.

Sometimes a brief anecdote becomes a perfect springboard to a new area of study. The story of Galileo, for example, is a useful way to introduce the development of the scientific method. The metric system makes more sense when one has heard the story of Lagrange and the chaos in Europe when each city had its own system of weights and measures. Why do we use *a, b,* and *c* to label the sides of a triangle? Well, let me tell you about Euler. Students will be more interested and more likely to remember what they study when they connect mathematical truth with real persons and practical situations.

Problem-Solving Savvy

The historical perspective accentuates many different approaches to problem solving. One could teach a unit on multiplication, including lattice multiplication (fig. 10.1), multiplying with Napier's rods (fig. 10.2), and the Russian peasant method. When students learn by their own experiences that products obtained using these different techniques match the products obtained from the most common algorithm, they discover that mathematics is not restricted to only one way of solving problems. Mathematics thus takes on an international, multicultural flavor.

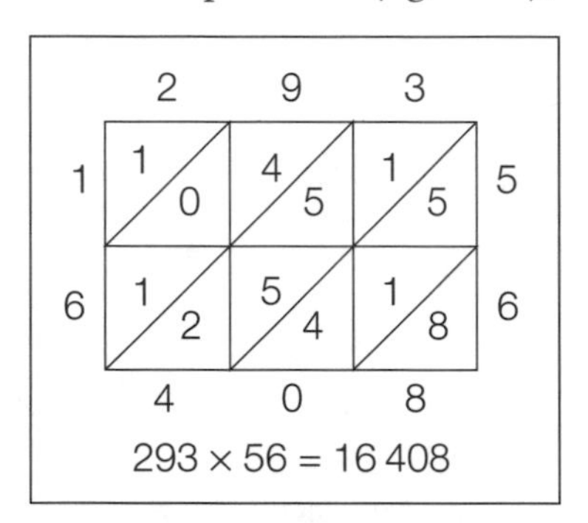

Fig. 10.1. Lattice multiplication

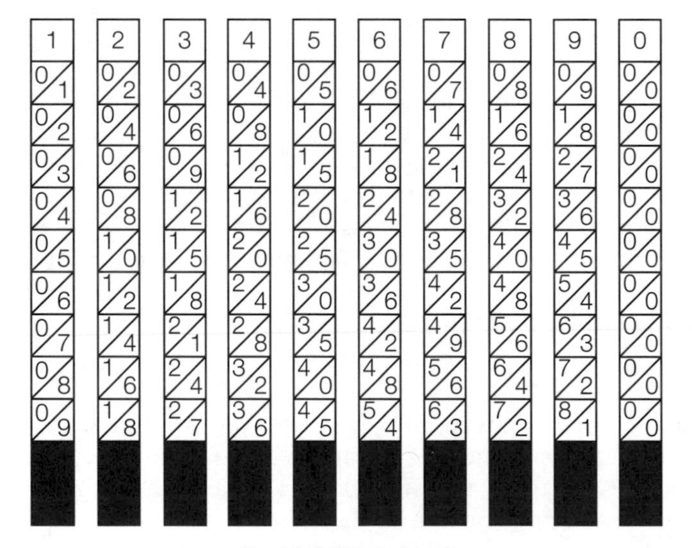

Fig. 10.2. Napier's rods

Mathematical Linkages

Exploring historical origins often opens a window to the interconnectedness of mathematics itself. Descartes's revolutionary combination of algebra and geometry led to a breakthrough in mathematical understanding. The story of how he applied one to the other to create analytical geometry shows students that one mathematical concept often enhances and strengthens another in a remarkable way.

Connecting with Today

Finally, mathematics history creates a bridge from the past to the future. While still a boy, Blaise Pascal was determined to design a machine to help his overworked accountant father. It took him several years to perfect his invention, but finally he succeeded. Today, that machine is considered the first mechanical calculator. For this reason, a popular computer language is named Pascal.

In 1993, newspapers announced that "Fermat's Last Theorem," a problem that has intrigued mathematicians for hundreds of years, had at last been solved. Whether the proof was flawed or not, the headlines reminded the world that mathematics is not dead but still alive and developing.

HOW CAN HISTORICAL PERSPECTIVES BE TAUGHT IN MATHEMATICS?

An Integral, Not Additional, Dimension

The NCTM Standards stress that all students, including those with minimal motivation, can learn to value mathematics more by being exposed to its historical and cultural contexts. Teachers agree in principle, but many are concerned about adding yet another unit to an already crowded curriculum.

Including a historical dimension in the mathematics classroom does not mean replacing any part of the mathematics curriculum. An awareness that mathematics *has* a history should permeate the entire teaching environment. Instead of saving history as a "filler" for rainy days or reserving it as "enrichment" for advanced students, it should be integrated and used on a daily basis. Teachers must demonstrate that doing mathematics is a part of being human.

Creating a Historical Environment

One very simple way to introduce historical elements is through classroom displays. Time lines visually depict the development of mathematics and help students place achievements in perspective with other historical events. Portraits of great mathematicians with brief biographical notes encourage familiarity; for instance:

Galileo Galilei
1564–1642

> "Mathematics is the alphabet with which God has written the universe."
> Galileo

Some of the world's greatest wisdom has originated with mathematicians. Classroom displays of quotations from mathematicians is another effective way to stimulate an appreciation for the breadth of their contributions. Consider how this quote from Isaac Newton might provoke discussion (Bell 1937, p. 93): "If I have seen farther than others, it is because I have stood on the shoulders of giants."

Some mathematicians have also been immortalized in busts or small statues suitable for classroom display. Simply having one on a shelf in the classroom sends a signal to students: in this room we value mathematicians!

Make It Stick

Stamp collecting is a popular hobby with young people, and it offers another natural connection to mathematics history. Teachers and students may develop special albums to display stamps honoring mathematicians and their discoveries. This quickly becomes a mini-lesson in geography, history, language, and culture as stamps from all around the world are collected.

Carl Friedrich Gauss — Isaac Newton

Reprinted from Schaaf (1978) with permission

Reading Is Recommended

Many teachers are wisely trying to integrate more reading and writing into their mathematics teaching. Mathematics history presents a perfect context for this. Students respond enthusiastically to simply hearing stories read aloud. Uncopyrighted illustrations may be enlarged and copied onto transparencies for display at appropriate times during the reading. School and classroom libraries should contain a range of materials, including biographies, for students to read independently.

Writing Is Right!

Many options for writing projects arise from mathematics history. Students can research the life of a particular mathematician and write a report. They might read a biography or a historical novel about a mathematician and write a book review. Teachers may suggest they write an imaginary interview, a newspaper story, a screenplay, a poem, or song lyrics about an individual or a discovery from mathematics history. Other possibilities include writing about the origin of a particular concept or symbol, such as π or $=$. Some might wish to write about how mathematics was understood in a particular place or period.

Get Dramatic!

Mathematics history adapts very well to the stage. To prepare, students may read or listen to a brief biographical sketch or a collection of anecdotes about a selected mathematician. Older children may do library research. Small groups of students may then be invited to prepare skits or, if equipment is available, videos to share with the class. Students could stage a newscast, an interview, or a *This Is Your Life* show. Memorizing lines, using costumes and props, and selling tickets are, of course, optional.

Through the Arts

Mathematical history projects also reinforce the visual arts. Students may draw or paint a scene from the life of a mathematician. They may need to be reminded of the importance of researching the architecture, clothing, and furnishings of the time and place in which the mathematician lived.

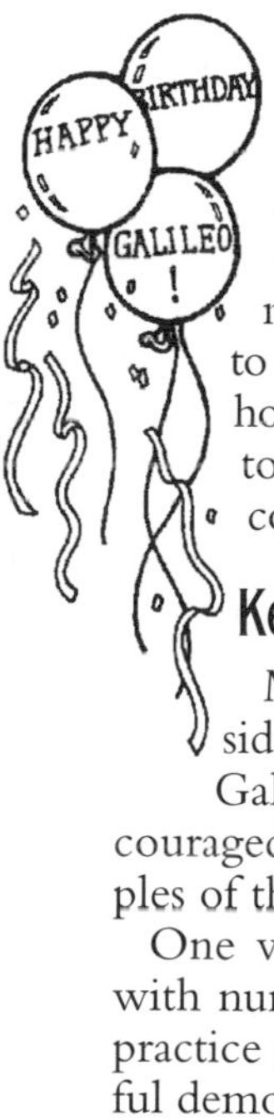

HAPPY BIRTHDAY GALILEO !

Happy Birthday!

A wide range of connecting ideas may culminate in a classroom birthday party for a particular mathematician. Some teachers keep a calendar on display all year long to highlight mathematicians' birthdays. Groups of students may decorate a bulletin board in honor of the mathematicians who have birthdays that month. A teacher may choose to highlight the birthday with a simple recognition of the date and the honored mathematician's contributions. Others may use the occasion to read a brief biography. Still others may prefer a full-scale party, complete with cake and party games related to the mathematician.

Keep It Hands On

Many historical activities promote hands-on experiences. Consider arithmetic on the abacus, multiplication using Napier's rods, or Galileo's experiment with falling objects. Students should be encouraged to conduct experiments or do surveys, reinforcing the principles of the scientific method as a problem-solving tool.

One very simple but effective activity connects to Pythagoras's work with numbers as shapes (Reimer and Reimer 1992). Students use plastic practice golf balls and a glue gun to construct number models for powerful demonstrations of number relationships (fig. 10.3).

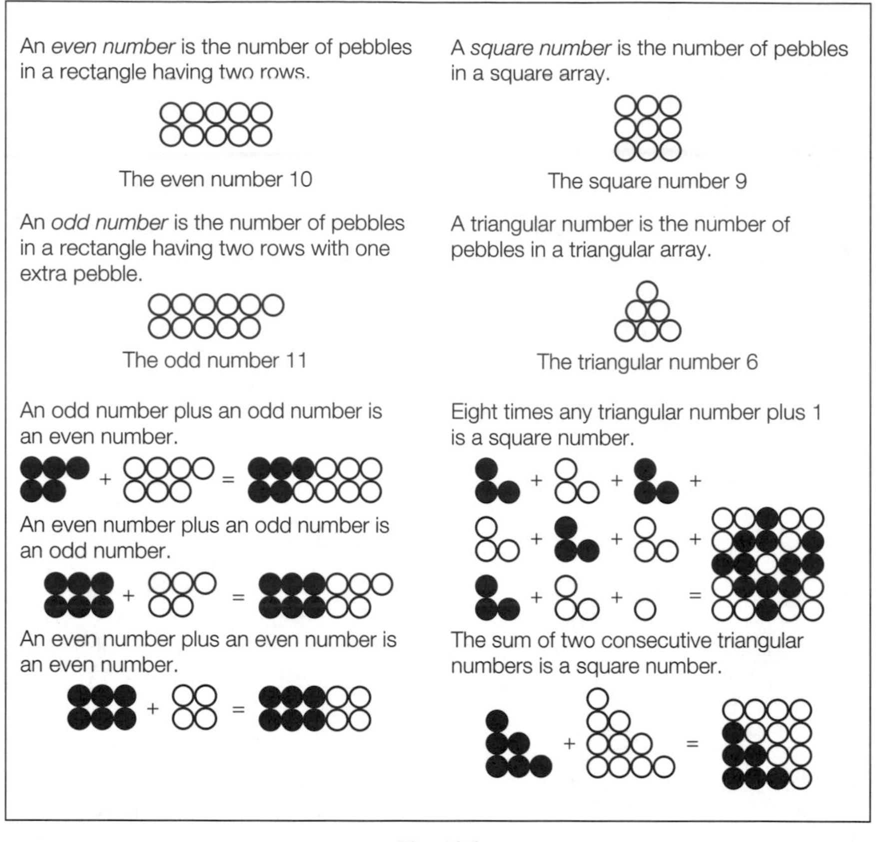

Fig. 10.3

Retracing Their Steps

Although Pythagoras would have used pebbles instead of practice golf balls, leading students to make discoveries and solve problems in the footsteps of the great mathematicians is perhaps the most natural and productive way of integrating historical elements into the mathematics classroom. Activities may be used that either replicate or parallel the problems faced by influential mathematicians from the past.

Take Gauss, for example. Tell your students the story of how, as young boys, Gauss and his classmates were given the onerous task of summing the first 100 counting numbers. Gauss found an easy way to do it. Can your students? Similar problems ripple out from this one. How would students recommend summing the first 100 even numbers, or the first 100 odd numbers?

Although Sonya Kovalevsky's work in infinite sequences may be too complex for some students, an exploration of the snowflake curve will fascinate them. Here is an additional opportunity to include a technological component, a simple Logo program that enables students to watch the perimeter become "infinitely large" while the area approaches a finite quantity (figs. 10.4 and 10.5).

```
TO SIDE :SIZE :STAGE
IF :STAGE = 1 [FD :SIZE STOP]
SIDE :SIZE/3 :STAGE -1
LT 60
SIDE :SIZE/3 :STAGE -1
RT 120
SIDE :SIZE/3 :STAGE -1
LT 60
SIDE :SIZE/3 :STAGE -1
END

TO SN :STAGE
MAKE "SIZE 180
PU SETPOS [-50 -75]
PD
REPEAT 3[SIDE :SIZE :STAGE RT 120]
END
```

To run the program, enter SN followed by a number that indicates the stage to be drawn. For example, when SN 1 is entered, the turtle should draw stage 1 of the snowflake curve. When SN 2 is entered, stage 2 is drawn, and so on. Draw a number of stages, clearing the screen (CS) between drawings.

Fig. 10.4. Logo program for the snowflake curve

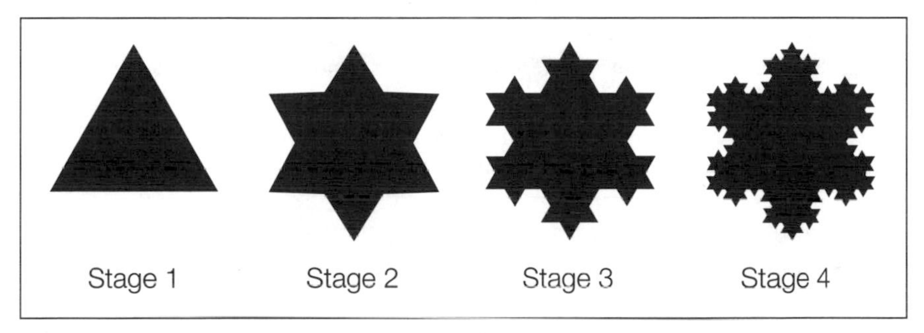

Fig. 10.5

Satisfaction Guaranteed

Connecting mathematics with its history makes learning the basic skills more interesting and motivates students to sustain their interest in mathematics. With a little effort, teachers will be able to spot appropriate places to include historical materials in their lesson plans. Resources available now easily link concepts with personalities and place procedures into historical context. As many have reported, interest in mathematics increases not only for students but also for teachers.

Connecting mathematics to its history does not have to happen only during mathematics class. Because the historical elements interface so beautifully with social studies, reading, writing, and art, a holistic approach is not only possible but ideal. Reinforced learning is lasting learning!

RESOURCES IN MATHEMATICS HISTORY FOR LIBRARY AND CLASSROOM

The following annotated list may serve as a starting point from which to choose resources about the history of mathematics for your school library or classroom:

Abbott, David. *The Biographical Dictionary of Scientists: Mathematicians.* New York: Peter Bedrick Books, 1986.

This is an authoritative and accessible reference work. It includes a chronological introduction and alphabetical arrangement of entries, plus a useful glossary.

Beckman, Petr. *A History of Pi.* New York: St. Martin's Press, 1971.

A readable, interesting source that reveals the background of the times and the personalities associated with the development of pi.

Bedini, Silvio A. *The Life of Benjamin Banneker.* Rancho Cordova, Calif.: Landmark Enterprises, 1972.

A definitive biography of the first African American of science.

Bell, Eric Temple. *Men of Mathematics.* New York: Simon & Schuster, 1965.

This classic work in mathematics history includes extensive detail and useful information.

Dunham, William. *Journey through Genius: The Great Theorems of Mathematics.* New York: John Wiley & Sons, 1990.

This work explores some of the most significant and enduring ideas in mathematics, emphasizing the humanity of the great mathematicians.

Edeen, Susan, and John Edeen. *Portraits for Classroom Bulletin Boards.* Books 1 and 2. Palo Alto, Calif.: Dale Seymour Publications, 1988.

Both of these books contain fifteen line-art portraits and biographical sketches, which may be reproduced for bulletin-board or classroom use.

Eves, Howard W. *An Introduction to the History of Mathematics.* 6th ed. New York: Saunders College Publishing, 1990.

A popular text for history of mathematics classes. Eves traces the development of mathematics with good humor and informative detail.

———. *In Mathematical Circles.* Vols. 1 and 2. Boston: Prindle, Weber & Schmidt, 1969.

These popular books contain chronologically arranged anecdotes about mathematicians and their discoveries as well as delightful, short bits of useful and understandable information.

Ipsen, David C. *Archimedes: Greatest Scientist of the Ancient World.* Hillside, N.J.: Enslow Publishers, 1988.

This book, written for young people, traces the life and discoveries of the Greek mathematician, scientist, and inventor.

———. *Isaac Newton: Reluctant Genius.* Hillside, N.J.: Enslow Publishers, 1985.

A biography of the seventeenth-century English scientist who developed the theory of gravity, discovered the secret of light and color, and formulated the system of calculus. Written for juvenile readers, the book includes many helpful drawings and illustrations.

Mitchell, Merle. *Mathematical History: Activities, Puzzles, Stories, and Games.* Reston, Va.: National Council of Teachers of Mathematics, 1978.

A collection of enrichment resources for use in the upper elementary grades. Activities may be photocopied for classroom use.

Multiculturalism in Mathematics, Science, and Technology. Reading, Mass.: Addison-Wesley Publishing Co., 1993.

Designed to help infuse multicultural education into science and mathematics classrooms. More than fifty activity-oriented lessons are included.

National Council of Teachers of Mathematics. *Historical Topics for the Mathematics Classroom.* Reston, Va.: The Council, 1989.

This work, first commissioned in 1969 as the NCTM's Thirty-first Yearbook, is designed to help teachers teach mathematics from a historical perspective. It is divided into chapters on the history of numbers, the history of geometry, the history of algebra, and so on, and includes a useful list of resources.

Pappas, Theoni. *The Joy of Mathematics.* San Carlos, Calif.: Wide World Publishing/Tetra, 1989.

Unveils the inseparable relationship between mathematics and the world in which we live. In one- or two-page "glimpses," the reader enjoys games, puzzles, interesting facts, and historic background.

———. *More Joy of Mathematics.* San Carlos, Calif.: Wide World Publishing/Tetra, 1991.

Like Pappas's first book, this collection also provides brief but fascinating information on how mathematics can be seen in nature, science, music, architecture, literature, and history.

Perl, Teri. *Math Equals: Biographies of Women Mathematicians.* Reading, Mass.: Addison-Wesley Publishing Co., 1978.

A readable collection of resources on the lives and work of nine women, including activities that relate to their work.

Reimer, Luetta, and Wilbert Reimer. *Mathematicians Are People, Too.* Vols. 1 and 2. Palo Alto, Calif.: Dale Seymour Publications, 1990, 1994.

These collections of illustrated stories dramatically re create episodes from the lives of fifteen mathematicians, including three women. For students to read or for teachers to read aloud, the books highlight the human element in mathematics. Appropriate for students in grade 3 through secondary school.

———. *Historical Connections in Mathematics: Resources for Using History of Mathematics in the Classroom.* Vols. 1, 2, and 3. Fresno, Calif.: AIMS Education Foundation, 1992, 1993, 1994.

Designed to facilitate the linkage between history and mathematics, these collections each include biographical information, famous quotations, fascinating anecdotes, and more than eighty illustrations from the lives of ten mathematicians. Each chapter features reproducible activities (more than fifty in all) related to the mathematician's work. Suggestions and complete solutions are included.

Schaaf, William. *Mathematics and Science: An Adventure in Postage Stamps.* Reston, Va.: National Council of Teachers of Mathematics, 1978.

Through illustration and historical insight, this book traces the way postage stamps mirror the impact of mathematics and science on society.

Zaslavsky, Claudia. *Africa Counts.* Brooklyn, N.Y.: Lawrence Hill Books, 1973.

Describes the contribution of African peoples to the science of mathematics using pattern and number as organizing principles.

REFERENCES

Bell, Eric Temple. *Men of Mathematics.* New York: Simon & Schuster, 1965.

Reimer, Luetta, and Wilbert Reimer. *Historical Connections in Mathematics: Resources for Using History of Mathematics in the Classroom.* Vol. 1. Fresno, Calif.: AIMS Education Foundation, 1992.

Schaaf, William. *Mathematics and Science: An Adventure in Postage Stamps.* Reston, Va.: National Council of Teachers of Mathematics, 1978.

PART 3

Connections across the Elementary School Curriculum

11

Learning Mathematics in Meaningful Contexts: An Action-Based Approach in the Primary Grades

Sydney L. Schwartz
Frances R. Curcio

YOUNG children begin school with a rich informal knowledge base, an extensive repertoire of skills, and a wide variety of interests. Their curiosity and enthusiasm for exploring the world around them include a pervasive use of emerging mathematics concepts and skills. For the past three decades, schools have increasingly sought to capitalize on children's natural inclinations as active learners by providing materials-rich classroom environments for kindergarten and primary grades (Bruner, Jolly, and Sylva 1976; Wasserman 1990).

Bringing mathematics alive for young children in school requires activities that converge on the needs for *meaningful context* and for the *integrity of mathematics content.* Kindergarten programs have generally been successful in creating an action-based, learning-centered environment, but the mathematics curriculum has often been invisible both to teachers and to children. Many primary school programs have supplied manipulative materials as props for completing paper-and-pencil tasks, but the use of the materials has primarily served the purposes of decontextualized drill. The challenge remains to design activities that conform with children's interests, stimulate mathematics learning in context, and meet school concerns for mastering the curriculum.

One of the most promising vehicles to accomplish these aims is an integrated curriculum model. The variety of activities in this approach gives children an opportunity to make connections between observed objects and events and the abstract ideas that explain relationships among objects and events. These activities include investigating with concrete materials; reading literature and recording information through narratives, pictures, graphs, and charts; analyzing, interpreting, and reporting data; and playing games. The power of this action-based learning approach is that activities grow out of children's interests instead of being externally imposed as isolated tasks. Further, the activities can more effectively connect with individual developmental needs, allowing children to build on what they know and can do (Schwartz and Robison 1982). Interest-driven activities offer learning sequences that draw on the natural learning patterns of children (Wasserman 1990; Monighan-Nourot, Scales, and Van Hoorn 1987).

This article presents a segment of a unit on animal habitats developed by a teacher, Ms. V, who built on children's interests. The major activity for this part of the unit was collecting materials and constructing a bird's nest. It was sparked by one child's report of sighting a bird's nest. As children engaged in the curriculum activity that evolved, they dealt with concepts and skills across the curriculum: mathematics concepts through creating equivalent sets and exploring number relationships in the process of solving problems; science concepts through studying the conditions under which birds construct nests and the properties of nesting materials; communication arts through reading, writing, and talking about their experiences; the creation of models through designing and constructing a bird's nest; and social studies concepts through comparing people's and animals' dwelling places. To capture the continuity of the dialogue between the teacher and the children in a first-grade class, the mathematics segment of the following vignette was excerpted from a videotape. The teacher chose to have the children work in small groups so that each child could be an active participant rather than a passive observer.

BUILDING A BIRD'S NEST

During a first-grade sharing session, Jeanette excitedly described a bird's nest that she had seen outside her kitchen window. The children asked her questions: "How many birds live in the nest?" "What does the nest look like?" "How high is it in the tree?" They expressed interest in how birds make nests.

The following day, Ms. V brought a bird's nest to class. As the children examined it, they wondered whether the materials in this nest could be found in their neighborhood. In response to their interest, the teacher and children planned a neighborhood walk.

Several days later, the class went on a nature walk searching, as birds do, for materials they could use to make nests. Working in groups of four, the children found buttons, twigs, yarn, grass, and leaves, placing their findings in large plastic shopping bags.

When the children returned to class, Ms. V posed a question: "What should we do with this material so that everyone will get a fair share of everything to make a nest?" Tony suggested that they put all similar objects together into groups. Avril added that if everyone did this, the piles of things that are alike could be combined. Everyone agreed and dumped the materials to be sorted onto their desks.

After the sorting and compiling was completed, each group of four to six children took responsibility for one set of materials. Ms. V then began posing problems to facilitate the next step in sharing the materials: "We need to know if everyone in the class can have a button. What should we do to find out?"

Suzette suggested, "Count them."

Ms. V asked the children to continue to solve the problem of sharing the objects. Children visited other groups to find out what strategies they were using. As the groups completed their tasks, the teacher initiated a discussion about how they generated their solutions.

Jackie, Jeanette, Jeffrey, John, Suzette, and Avril worked together figuring out how to distribute the buttons fairly. They counted the buttons—88 in all. As the children worked, Ms. V observed them and periodically engaged them in discussion on their progress and next steps.

Ms. V: Now that you know you have 88 buttons, what other information do we need to figure out this problem?

Jackie: We have to see if we can get 23 or two 23s, or any amount of 23s for all the kids.

Ms. V: Why did you pick 23?

Jackie: Because of 23 kids in the class. We have to split it for all the kids.

Ms. V: How can you do this?

Jeffrey: We can do it one by one.

Ms. V: So how many piles would you have?

Jackie: Twenty-three. Or we can make piles of 23.

Jeanette: We have 88 buttons and we have to split them up into 23 piles.

The children negotiated how to set up the buttons for distribution. They decided to make piles of 23. After making one pile, Jeanette said, "That's 88 minus 23," and wrote a number sentence to represent this. Jackie said, "Now we're gonna find another pile of 23." Jeanette then recorded, "That's 65 minus 23." Several children chimed in, "Two piles of 23." After they made the third pile, they again claimed, "Now we have three piles of 23." When attempting to make the fourth pile, however, they found they had only 19 buttons left, inspiring Jackie to assume a teaching role.

Jackie: If we had 19, Jeffrey, and we have to get to 23, how many more buttons do we need?

John: Five.

Jackie: I'm asking Jeffrey, but you were close. We have 19, right? How many more do we need to make 23?

The children worked on the problem, counting up on their fingers and finally agreeing that they needed four more buttons to make another pile of 23. At this point the teacher joined the discussion.

Ms. V: How many buttons will each child get?

Jeanette: Two.

Jackie: Three, because Sal would get one from this pile—that's one—and one from this pile, and that's two buttons, and one from this pile, that makes three buttons. If we get more buttons, then Sal can get four buttons.

Ms. V: How can we fix the materials so that each person could come to the table and just take his or her share of buttons?

With almost no discussion at all, the children changed their three piles of 23 buttons to 23 piles of three buttons. Jackie's written summary of the group's thinking was as follows:

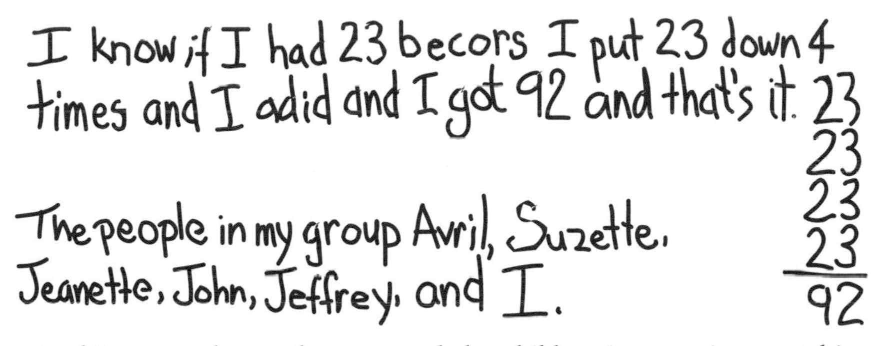

I know if I had 23 becors I put 23 down 4 times and I adid and I got 92 and that's it.

The people in my group Avril, Suzette, Jeanette, John, Jeffrey, and I.

23
23
23
23
92

In this event, the teacher engaged the children in counting, matching sets, and separating and partitioning sets into equivalent sets. Because of the collaborative nature of the task, the teacher could not monitor each child's mathematical participation. However, Ms. V was not distressed, since many of the activities in the program engage children in similar forms of mathematical behavior that promote interaction and challenge children's thinking about mathematical problems.

The vignette begs the question of what young children can do mathematically. These children clearly surpassed the standard expectations of a first-grade mathematics curriculum. However, they did it in the real-life world of mathematics rather than in the abstract, symbolic, written world of mathematics. Interestingly, a few of them were spontaneously crossing the bridge into the symbolic "world" without direct instruction. Through experiences like this one, teachers can begin to redefine direct instruction by supplying vibrant contexts in which mathematics is integral to the accomplishment of a goal. Teachers then can encourage children to use their own mathematics to find solutions in the context of the activity.

OTHER TOPICS TO EXPLORE

The following annotated list includes popular topics associated with certain themes. We are grateful to the many New York City early-childhood teachers who have designed and implemented variations of the ideas outlined in this annotated list. Each topic includes a statement of the context, connections across the curriculum, mathematics concepts, and a thumbnail sketch of how mathematical problem posing and problem solving can flourish. The suggestions for tasks range from simple to complex, offering possibilities for challenges that conform with the observed levels of understanding and skills of the participating children. Because of the desirability of involving all participants in thinking about solutions to problems, it is recommended that these activities take place with a small group of four to eight children.

Plants in Our Environment

Topic/Task: Sprouting seeds and bulbs

Context: During spring, children begin to notice the renewal of plant life. A variety of related activities may include taking walking trips in the neighborhood and local parks, planning the purchase of plant supplies, reading stories about gardening and farming, and talking with experienced gardeners.

Curriculum connections: Science, mathematics, communication arts

Mathematics concepts: Recording data, measuring, counting, joining, separating, multiplying, and dividing (partitive)

Activity: Initially, children figure out how to distribute materials for the planting project: *How many seeds will each person get if everyone gets the same number?* A more complicated problem combines multiplication and partitive division: *If we wanted to have 5 seeds for every child and there are 20 seeds in each package, how many packages will we need to buy?*

To measure and record plant growth, children select measurement tools from such possibilities as a single nonstandard unit to describe length (e.g., a pipe cleaner or strip of paper); a set of nonstandard units (e.g., Unifix cubes); standard units (e.g., centimeter or inch cubes); or a ruler.

Recording their observations in pictures, symbols, charts, and graphs establishes the basis for thinking about relationships between elapsed time and plant growth: *What is the difference in size between these two plants? How many units taller is the plant today than it was last Monday? How can you figure out whether the plant has grown the same amount each week? How much do you think the plant will grow between now and next Friday?*

Favorite Foods

Topic/Task: Making applesauce

Context: Cooking projects fit naturally into a variety of units and themes: multicultural themes, food production, restaurants, special events, properties of objects, and nutrition. Children frequently discuss their food likes and dislikes.

Curriculum connections: Science, social studies, mathematics, communication arts

Mathematics concepts: Numerical comparisons, multiplication, division, and fractions

Activity: Children help determine the amount of ingredients. Initially, children figure out the amount of ingredients needed on the basis of the number of eaters. *If every person will be eating two cooked apples, how many apples do we need to cook?* Later: *This recipe requires two cups of sugar. We have only half-cup measures to use. How many half-cup measures of sugar will we need to put into the mixing bowl to get two cups of sugar?*

Additional problems evolve from procedures for serving food and talking about other attributes, such as the number of seeds: *Do you think we will find the same number of seeds in each apple? How many seeds will there be altogether if there are X seeds in each apple and we are cooking 20 apples?*

Buyers and Sellers

Topics/Task: Classroom store for classroom project materials

Context: The children frequently discuss shopping experiences and initiate playing store.

Curriculum connections: Economics, social studies, mathematics, communication arts

Mathematics concepts: Joining, separating, comparing, and multiplying

Activity: The teacher and the children set up a supply store for "purchasing" materials to complete craft projects, using play money. Each child receives a specified amount of play money in coin denominations appropriate to the level of the class. All art materials, for example, are priced at one, two, or five cents a unit. Two children collaborate in the role of storekeeper, helping each other figure cost and change due as they serve each customer. The "customers" answer questions about their purchases: *Your purchases cost six cents and you gave the two storekeepers ten cents. How much money will you get back as change? If you were going to make two more place mats just like this one, how much more money would you need to give the store clerk?*

Mapping

Topic/Task: Writing directions

Context: A plan to invite visitors to the classroom requires written directions for traversing the school building.

Curriculum connections: Geography, mathematics, communication arts

Mathematics concepts: Measuring, counting, specifying directionality, and translating measurement information from three-dimensional settings to two-dimensional representation

Activity: Children figure out how many pathways there are between the classroom (e.g., on the second floor at the rear of the building), and the front door of the school building. They locate important points such as the ends of corridors and figure out distances between these points. Then they translate this information into written directions.

Animals

Topic/Task: Solving animal riddles

Context: Children learn about how animals live.

Curriculum connections: Science, social studies, art, creative writing, mathematics

Mathematics concepts: Joining, separating, comparing, and nonroutine problem solving

Activity: The teacher poses riddles that call on children's knowledge of animal movement and appendages. Simple problems require selecting two animals with the same number of legs. For example: *There were two animals in the lake. There are a total of four legs. What could the animals be?* The children are encouraged to discuss possible solutions with one another. They soon discover that more than one solution is possible, such as one boy (two legs) and one girl (two legs), or one dog (four legs) and one fish (no legs).

More complex problems call for thinking about an increased number of animals or legs. *There were three animals in the lake and a total of eight legs. What could the animals be?* Later, the teacher invites the children to make up riddles for their classmates. A riddle posted on the bulletin board furnishes additional opportunities for children to solve problems and post solutions. The teacher and the children check responses and compare for similarities and differences in solutions.

Daily Routines

Task: Distribution of snack materials

Context: Although routines are not usually considered part of the academic curriculum in the kindergarten and primary grades, many problem-solving opportunities emerge from everyday classroom activities and daily routines. The distribution of materials at snack time and during many of the program activities can furnish a context for solving mathematics problems. The solution to these problems not only facilitates the management of the classroom activity in progress but also fosters the development of the children's ability to manage routines independently.

Connections: Social living skills, mathematics, communication arts

Mathematics concepts: Counting, joining, separating, multiplying, and dividing

Activity: Each day, as children fulfill responsibilities for distributing napkins, straws, milk containers, and crackers, the teacher poses word problems to the table servers. *You are going to give napkins to everybody at your table and at Sal's table. How many napkins will you need?* When setting up the cracker tray for one group of children, the teacher queries: *This table has three children and each child will get two crackers. How many crackers should I put in the basket for this table? There are four children at this table and I am putting eight crackers in the basket. How many crackers should each child take so that everyone will have the same number?*

REFERENCES

Bruner, Jerome, Alison Jolly, and Kathy Sylva, eds. *Play—Its Role in Development and Evolution.* New York: Basic Books, 1976.

Monighan-Nourot, Patricia, Barbara Scales, and Judith Van Hoorn. *Looking at Children's Play: A Bridge between Theory and Practice.* New York: Teachers College Press, 1987.

Schwartz, Sydney L., and Helen F. Robison. *Designing Curriculum for Early Childhood.* Boston: Allyn & Bacon, 1982.

Wasserman, Selma. *Serious Players in the Primary Classroom.* New York: Teachers College Press, 1990.

12

Measurement in a Primary-Grade Integrated Curriculum

Lynn Rhone

"Mathematics as Communication," Standard 2 from NCTM's *Curriculum and Evaluation Standards for School Mathematics* (1989, p. 26), focused my thoughts about creating experiences for young learners as they explore mathematics:

> Young children learn language through verbal communication; it is important, therefore, to provide opportunities for them to "talk mathematics." Interacting with classmates helps children construct knowledge, learn other ways to think about ideas, and clarify their own thinking.

This year I have a multiaged kindergarten, first-grade, and second-grade classroom with twenty-five children; seventeen first and second graders all day, with eight kindergartners joining us in the afternoon. My current second graders have been with me for three years, so it has been extremely interesting to watch the development of their thinking and reasoning in mathematics. Next year my first graders and kindergartners will be joining me for a second year. Once again I'll be able to watch their development and to challenge them in many directions. The multiage philosophy at my school is to allow children to stay with the same team of teachers for three years for the primary years and to go on to intermediate grades 3, 4, and 5 for another three years.

The children in my multiaged primary school classroom have been engaged in activities built around integrated-curriculum units throughout the year. They have participated in a kaleidoscope of activities using manipulative materials; the language arts strands of speaking, listening, reading, and writing; the science and social studies processes; and the creative arts. As I designed lessons for this integrated curriculum, I saw a pattern emerge, a plan that could be used with each lesson. The plan has the following six steps:

1. Identify the mathematics and language arts concepts or objectives.
2. Plan the performance tasks linking the instruction and activity with the performance assessment.
3. Choose and read a story or poem related to the integrated-curriculum theme or to the mathematics objective.

4. Allow children time to work on the mathematics objective using appropriate manipulatives.
5. Allow children time to record their reasoning in pictorial and written form.
6. Allow children time to share their reasoning at the "math author's chair," a special chair for students to explain to their classmates their own solution to a problem.

The activities described here covered three days and depict primary-aged children exploring many ideas and concepts. Steps 1, 2, and 3 introduced the activities and the assessment rubric; steps 4, 5, and 6, which were revisited each day, describe those activities in detail. The language arts strands of speaking, recording pictorially, and writing enabled the children to explain their reasoning.

THE BEGINNING STEPS

Step 1. The children had engaged in many activities about measurement during the year. This particular series of lessons took place at the end of the year and combined linear measurement, place value, graphing, and averaging. The mathematics objectives were integrated with the language arts strands of speaking, recording pictorially, and writing.

Step 2. Assessment, when closely planned and linked within the instruction, focuses the instruction and benefits the students. The performance-assessment task for this unit was planned to be a natural part of the learning process. The children knew what was being asked of them, and they were shown the assessment rubric that would be used (fig. 12.1). It is intended to be used over three years and is designed to show growth for each child.

Step 3. Earlier in the year the children and I had read Syd Hoff's *Lengthy* (1964). Much discussion of height and length had followed. Now, as we discussed height and length again, I read them Leo Lionni's story, *Inch by Inch* (1960). The story involves an inchworm who measured many objects in nature; the story thus focused our attention on standard units of measurement. Although we had used nonstandard units of measure earlier in the year (e.g., Unifix cubes, colored tiles, pieces of string), we now moved to English units of measure for comparison, inches in particular. I talked with the children about inches, and most of them remembered that inches measure things. I explained that I was going to help them find their heights in inches.

MEASURING HEIGHT IN INCHES

Step 4. I asked the children how long the worm in the story was. All the children knew it was an inch long because it was an *inchworm.* I showed the children a ruler and a yardstick and explained that those tools are used to measure inches. I also explained that our colored tiles could be used to measure

Mathematics and Language Arts Assessment Tasks—Grades K–2

Name ______________________________ Date ______________

Statement of Problem or Task: After reading *Inch by Inch* (Lionni 1960), children measured their heights using Unifix cubes and compared that measurement with their height in inches. Next, they counted cubes and inches by grouping by tens. Then they helped make a class graph of heights and participated in finding the average height of three classmates. Children spoke, recorded pictorially, and wrote their reasoning and results.

Mathematics Concepts:

- Comparing nonstandard and standard units of measurement
- Grouping and counting by tens
- Making a class graph of heights
- Finding an average height

Language Arts Concepts:

- Speaking at the math author's chair
- Recording pictorially and in writing

Mathematics Rubric: The child—

4

- accurately compares, measures, and counts cubes and inches;
- accurately groups and counts by tens;
- accurately helps make a class graph of heights;
- participates in finding the average height.

3

- measures and counts cubes and inches;
- groups and counts by tens;
- helps make a class graph of heights;
- participates in finding the average height.

2

- measures and counts cubes and inches inaccurately;
- groups and counts by tens inaccurately;
- inconsistently helps make a class graph of heights;
- inconsistently participates in finding the average height.

1

- needs help in measuring and counting cubes and inches;
- needs help in grouping and counting by tens;
- needs assistance to help make a class graph of heights;
- needs help to participate in finding the average height.

0

- does not attempt task.

Language Arts Rubric: The child—

4

- confidently and clearly speaks at the math author's chair;
- confidently and clearly records pictorially and writes.

3

- speaks at the math author's chair;
- records pictorially and writes.

2

- does not speak clearly and confidently at the math author's chair;
- does not record pictorially or write clearly and confidently.

1

- needs help to speak at the math author's chair;
- needs help to record pictorially and to write.

0

- does not attempt task.

Fig. 12.1

inches. Together we matched 12 colored tiles with the ruler and 36 colored tiles with the yardstick. To count the total number of inches, we put the colored tiles into groups of tens. We found there were four groups of tens and eight more, or 48 colored tiles, and thus 48 inches.

I then asked the children how we could use the tools to measure heights. One child said that at home his dad measured how tall he was by marking his height on a wall. He further explained that his family used a yardstick to tell how tall he was. We decided to tape the ruler and the yardstick to the wall and to measure everyone (fig. 12.2).

After we measured five children, we found that some were taller than 48 inches. Several children suggested that we add another ruler, so we then had two rulers and a yardstick. Beginning at 48 inches, we counted to 60 inches. As we measured each child, I cut a string that matched his or her height and attached masking tape to it to write the child's name and height.

I set out the Unifix cubes and asked the children to sit with a partner. Partners worked together to determine the number of cubes it would take to measure their lengths. The students invented several ways to do this. Most children lay on the floor while their partners made cube trains that matched their lengths (fig. 12.3). Other children stood while their

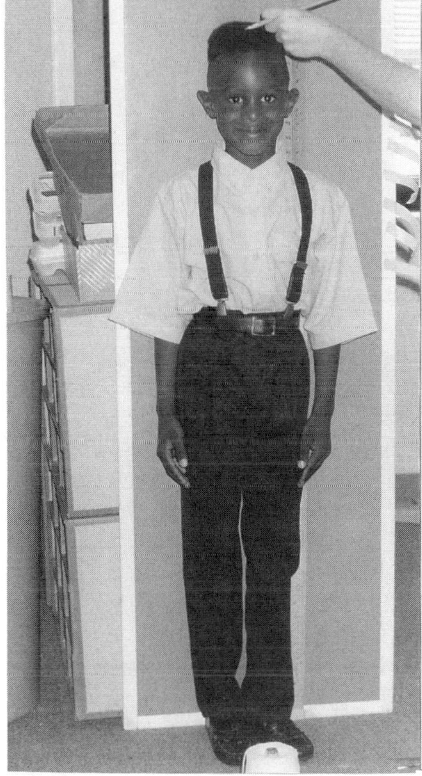

Fig. 12.2. Lamarr will see how many inches tall he is.

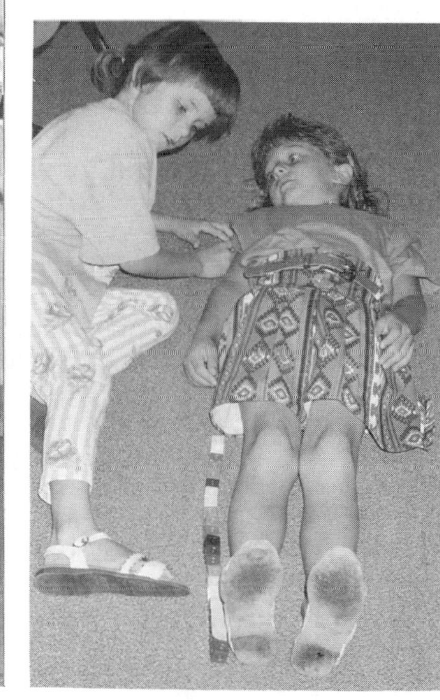

Fig. 12.3. Katie measures Brittany with Unifix cubes.

partners made the cube trains vertically; they discovered, however, that it was difficult to keep the cube trains together doing it that way. I took notes as they worked.

When they finished, each child had a cube train that approximated her or his length. I asked the children if they thought the cube trains would match the strings they had used to measure their heights earlier in inches. To find out, a member of the class stood while two children held the cube train next to him. It was amazing to them to find that the cube train and the string were not the same length. But when the string was held next to the same child, it was accurate. After discussing how they had erred in building their cube trains, the children put their strings next to their cube trains and added or took away cubes until the trains matched the lengths of the strings.

Step 5. I asked the children to think about how they were going to record their heights. We decided to use Unifix cube recording paper and to color the number of cubes used. I then asked them to cut out and paste their cube recordings on sheets of drawing paper, to draw pictures to represent their heights, and to write out the number of cubes in their height.

Step 6. At the math author's chair, the children were excited to show their portraits and the cube recordings of their heights. Already they were comparing their heights and talking about the tallest and the shortest in our class.

GROUPING BY TENS: PLACE VALUE

Step 4. The next day the children looked at the recordings they had made, and I asked them to make the trains again. I then asked the children to recall the number of inches in their heights, and I posed this question: "Will the number of cubes be the same as the number of inches?" The children thought that the two numbers would be the same.

Before the children started counting, we talked about the best way to answer the question. Several children thought they would count; others thought they would group by tens and then count. I said that I would be looking at how they were counting, and I would also be watching for them to group the cubes by tens. As the children worked, I recorded their progress.

When the children finished their counting, they noticed that the number of cubes was not the same as the number of inches, and yet the lengths were the same. Two children said they knew that Unifix cubes were not the same size as our colored tiles, so the numbers would be different. The children decided that it was important to say *what* units were used to measure, since the numbers would be different depending on the units of measure used (a difficult concept for the younger children in the group).

Step 5. I asked the children to record their findings on graph paper; I also asked them why they were going to use the graph paper. They discovered that the graph-paper squares matched the colored tiles. I explained that this graph paper was inch-square graph paper and that they were going to color the number of squares that showed how many inches tall they were (fig. 12.4).

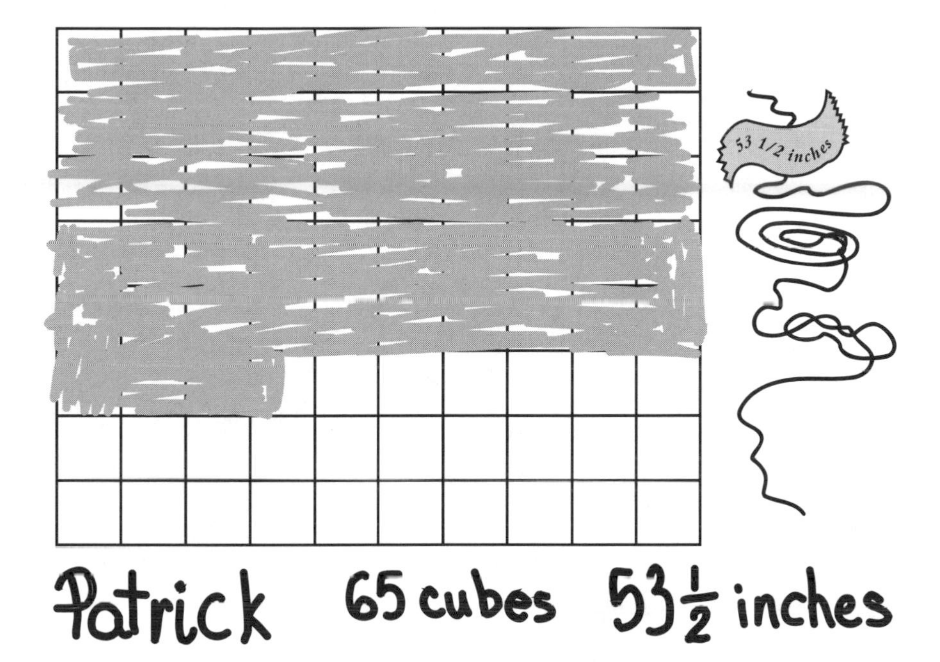

Fig. 12.4. Here Patrick uses inch graph paper to show his height as 53 1/2 inches. He is 65 Unifix cubes tall.

Before the children got their graph paper, we talked about how to show the number of inches accurately. One child recalled that when we counted the colored tiles and the Unifix cubes, we had grouped by tens, so he decided that he would cut strips of ten squares and color them. Another child said that since she was 42 inches tall, she would color four groups of ten squares blue and two more squares red. The children went to work as I recorded their progress.

Step 6. At the math author's chair, the children shared the number of inches tall that they were; they also told how many cubes tall they were. All of them found that the number of cubes was greater than the number of inches, but both numbers gave a way to record their heights.

MAKING A CLASS GRAPH OF HEIGHTS

Step 4. The next day, as the children came together, I gave them the strings that were used to measure their heights, and I asked them to tell me how we could use the strings to show the tallest children, the middle-sized children, and the smallest children in our class. In using the size words *small, middle-sized,* and *tall,* I wanted to see if anyone noticed the rhyming words *tall* and *small,* since we had done quite a bit of work with rhyming patterns. Katie was the first to say that *small* and *tall* rhyme.

One child put her string on the floor and said, "I am tall and so is Cailin." Cailin put her string on the floor. By that time, everyone had strings on the floor. I then asked, "What should we do next?" Matt suggested that we put all the long strings together, all the middle-sized strings together, and all the small strings together. After much discussion, the strings were placed in three groups. I told the children that they had done a fine job of grouping the strings. We now decided to make a graph so others could see the heights of all the children in our class. Using a large piece of paper, we labeled our graph Heights in Our Class.

Step 5. The three groups of strings were put in order according to height: the tall group, the middle-sized group, and the small group (fig. 12.5). Everyone thought the collection of strings served as a fine graph. I asked the children to draw a picture of the string graph and show where their height was on it. We talked about drawing the picture of oneself and how the writing could tell about one's size group. I recorded the way in which each child went about organizing his or her picture.

Step 6. At the math author's chair, the children decided to stand in a line from the tallest to the smallest. One child counted everyone in the line. Then we counted using ordinal numbers to find each student's place in our line.

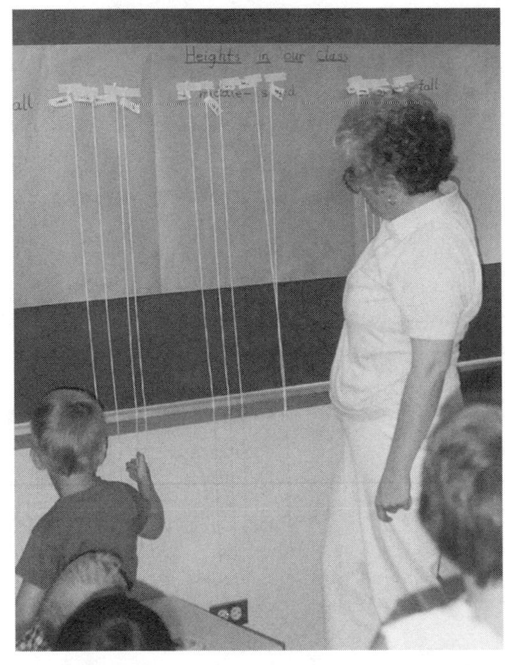

Fig. 12.5. We are finishing our class graph of heights.

FINDING THE AVERAGE HEIGHT OF THREE CHILDREN

Step 4. On the third day I reviewed the measurement activities we had completed over the previous two days. We talked about the string graph and how it showed the heights of all the students in our class. I explained that today we would find an average height for our class. As we began to talk about the meaning of *finding an average,* I explained that it would be a way to tell someone about the heights of the children in our class without having to see each height individually, as our graph shows.

Step 5. We had much discussion about what an average height is. We looked at the string graph showing each child's height. We chose the string of a tall child, one of a middle-sized child, and one of a small child and brought those strings to the floor. I asked the children to make cube trains to match the strings. We now had three trains.

I asked the children to tell me what they saw. They all agreed that they saw three trains of different lengths. I then asked them to consider making the three trains the same length without adding any more cubes or taking any away. They needed to rearrange the cubes in our selected trains so that all three trains would be the same length (or almost the same length). I explained that *finding the average* of the three heights, by using cubes, means making the three cube trains equal in length.

This was difficult to do, but we all worked together and made the trains appear as shown in figure 12.6. We talked about the cube trains and why they showed an average height of three children in our class. Several children noted that the trains now were about the same size because we rearranged the cube trains to make them even, and that the cube trains represented our three height groups. The cube trains showed an average height of 52 cubes.

Step 6. I wanted to find out what the children were thinking, so I asked them to draw pictures of our activity to find the average height. At the math author's chair the children talked about the average height that we found (fig. 12.7).

THE ASSESSMENT

The elements of the assessment were infused throughout the lessons, and assessment had a direct link to instruction. I constantly checked for understanding as the children worked, and I made notes about each child's progress.

In the first part of the project, when we measured length and height with cubes and then with inches, the children showed a wide range of abilities. Jamie was able to put her cubes into six groups of tens plus one more, making a total of 61 cubes. Patrick, the tallest member of our class, accurately counted and colored six groups of tens plus five more, making a total of 65 cubes. These children scored a 4 on the rubric for accurately counting cubes. Many of the younger children were not able to group by tens, so I helped them count the cubes by ones. They scored a 2 or a 1 on the rubric.

All the children were able to see a difference between the number of cubes and the number of inches; however, it was difficult for some children to talk about the size difference between inches and cubes. They needed more experience in measuring length with inches and cubes before they could truly understand it.

Fig. 12.6. Patrick "evens up" the cube trains to find the average.

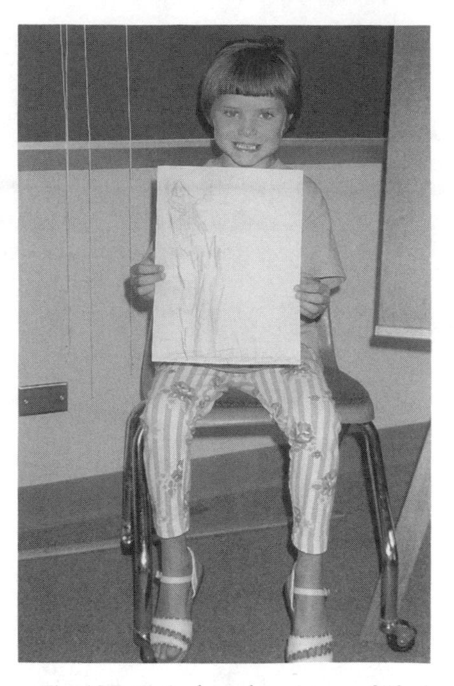

Fig. 12.7. Katie shares her picture of "finding the average" at the math author's chair.

In making the string graph showing the class's heights, the children worked well putting the strings into the *small, middle-sized, tall* groups. The children were eager to talk, draw, and write about their respective places on the graph. Lamarr drew a picture of himself and wrote, "I am in the tall GD," using an invented spelling for the word *group.* Leslie pictured herself and drew the measurement chart in the drawing. Matt eagerly shared his picture that showed him in the middle-sized group (fig. 12.8). Most children scored a 2, 3, or 4 on the rubric.

Finding the average height was an exciting activity for the children because it was the first time we had talked about averages. Lars and Patrick demonstrated great problem-solving skill by making the three cube trains even. We watched Patrick as he started the task by taking some cubes off the longest train and adding them to the two smaller trains one at a time. Then he said, "I'll add on five more to each train." He then had an equal number in each train, which showed an average height of 52 Unifix cubes. All the children scored a 2, 3, or 4 on the rubric.

For the language arts portion of the rubric, the children did extremely well at the math author's chair and scored 3s or 4s. Although the children do not write much, there were good verbal exchanges of ideas at the math author's chair.

Fig. 12.8. Matt says, "I am in the middle group."

Next school year, as I work with these children, we will have more opportunities to do activities in different settings. I look forward to giving the children time to talk about measuring and comparing in small groups and at the math author's chair. I plan to allow the children more time for group writing to record reasoning and to encourage more writing by them.

REFERENCES

Hoff, Syd. *Lengthy.* New York: G. P. Putnam's Sons, 1964.

Lionni, Leo. *Inch by Inch.* New York: Astor-Honor, 1960.

National Council of Teachers of Mathematics. *Curriculum and Evaluation Standards for School Mathematics.* Reston, Va.: The Council, 1989.

FOR FURTHER READING

Burns, Marilyn. *The Math Solution.* Sausalito, Calif.: Math Solution Publications, 1991.

Richardson, Kathy. *Developing Number Concepts Using Unifix Cubes.* Addison-Wesley Publishing Co., 1984.

Stenmark, Jean Kerr, ed. *Mathematics Assessment: Myths, Models, Good Questions, and Practical Suggestions.* Reston, Va.: National Council of Teachers of Mathematics, 1991.

13

Connecting Literature and Mathematics

David J. Whitin

AN IMPORTANT emphasis of the NCTM *Curriculum and Evaluation Standards for School Mathematics* (1989) is an urgent call to return context to mathematical learning. One effective strategy for restoring context to mathematical ideas is the use of children's literature (Whitin and Wilde 1992). Through books, learners see mathematics as a "common human activity" (NCTM 1989) that is used by people in different contexts for different purposes. This article describes how teachers of different grade levels have used the book *How Much Is a Million?* by David M. Schwartz (1985) to help children gain a deeper understanding of how mathematics is connected to the real world.

Raising Questions to Make Connections

Children sometimes have doubts about certain parts of Schwartz's story. Encouraging a skeptical attitude and supporting questions, however, help build a classroom of inquiring voices. Children raise questions about Schwartz's story in many ways. One example occurred in a fourth-grade classroom. When the teacher, Mr. Mingo, finished reading *How Much Is a Million?* to his students, he heard several of them ask, "Is that story really true?" Their response is typical for readers of this book in which the author tries to convey the enormity of one million, one billion, and one trillion by creating mathematical analogies. For instance, he claims that counting nonstop from one to one million would take twenty-three days; to count from one to one billion would take ninety-five years; and to count from one to one trillion would take more than 200 000 years. He uses other analogies as well: he describes how far one million children would stretch if they stood on each other's shoulders and how big a goldfish bowl would need to be to hold one million goldfish. He uses those contexts to describe the numbers one billion and one trillion as well. Thus, the statements in the book seem factual, but the analogies that are used, such as a large football stadium filled to the brim with water and holding one billion goldfish, clearly are fictitious. The question "Is this story really true?" certainly makes sense. Raising a question about an anomaly (i.e., about an idea that does not fit in with one's current understanding) is the hallmark of an inquiring learner. Questions are the essential catalyst for making connections and for reflecting on what we know.

Mr. Mingo was not quite sure what his students meant, and so he asked them to elaborate.

"You know, the part about taking twenty-three days to count to one million. Is that really true?" asked a student.

"What do you think?" asked Mr. Mingo.

"It just seems too long," another student replied. "It just can't take that long to count to one million."

"Well, how could you find out?" Mr. Mingo asked.

The children decided that they wanted to count to one million themselves. However, Mr. Mingo told them that the author had included some notes in the back of the book describing how he had arrived at his figures. For instance, Schwartz had based his calculation of twenty-three days on counting nonstop, twenty-four hours a day. "If he is right," Mr. Mingo asked, "do you want to stay at school that long in order to verify his figure?" The children decided against that suggestion but devised with their teacher a strategy that involved counting for sixty seconds.

Mr. Mingo selected five students to count for sixty seconds, each beginning at a different number. The children realized that some numbers would take longer to say than other numbers, so they tried to spread out the counting. The five children counted aloud several times so they could agree on a reasonable and consistent rate. Then they counted for one minute and used a calculator to determine what number they would reach if they kept counting for one hour and for one day (fig. 13.1). On the basis of Robert's rate,

	Per Minute	Per Hour	Per Day	Number of Days Required to Count at That Rate
Robert	1–85	5100	122 400	1 (to 122 400)
Leighann	1000–1043	2580	61 920	
Jaysen	10 000–10 037	2220	53 280	
LaTisha	900 000–900 031	1860	44 640	18 [122 400 + (44 640 • 18) = 925 920]
Jesse	900 950–900 970	1200	28 800	2 1/2 [925 920 + (28 200 • 2.5) = 1 000 000]

21 1/2 = Total number of days required to reach one million counting at Robert's rate for the first day, at LaTisha's rate for the next 18 days, and at Jesse's rate for the last 2 1/2 days

Fig. 13.1. Counting to one million

they would reach the number 122 400 by the end of the first day. Since their counting was now in the hundred thousands, they used LaTisha's rate to continue their counting; they used a calculator to add 44 640 for each day until they reached more than 900 000. Then they used Jesse's rate to figure another two and a half days' worth of counting. On the basis of their findings, they calculated that to count from one to one million would take almost 21.5 days. They were satisfied that their calculations were close enough to Schwartz's 23 days, and they surmised that the difference probably lay in the rate of counting. Thus, it was through this mathematical investigation that the children questioned the numerical information they had heard and as a consequence devised an experiment to test their hypothesis.

MAKING PERSONAL CONNECTIONS WITH MATHEMATICAL CONCEPTS

After reading aloud the entire story of *How Much Is a Million?* many teachers invite their students to create their own personal stories of one million. In one fourth-grade class, one student asked, "Do you think someone could drink a million gallons of water in a lifetime?" No one in the class was quite sure. At first they tried to determine how many gallons the average person drank in one day. Estimates ranged from one to ten gallons, but they finally decided that one-half of a gallon seemed like a reasonable figure. Then they calculated that it would take two days to drink one gallon of water, and therefore two million days to drink one million gallons of water. Next, they multiplied 85 years by 365 days to get 31 025 days. They realized at this point that even if they lived to age eighty-five, they would not be able to drink such a large amount of water.

Another exploration about water occurred in Ms. Grigg's sixth-grade classroom. One of her students, David, posed the following problem: "My dad's swimming pool holds 30 000 gallons of water. So how many pools would it take to hold 1 000 000 gallons?" The students used a calculator to find out that it would require about thirty-four pools to hold that much water. It is these kinds of personal comparisons that provide learners with benchmarks for understanding the magnitude of such large numbers.

Children in Ms. Bingham's second-grade class wanted to represent one million using parts of their body. Jessica suggested that they count noses; Alisha recommended arms; and John even suggested strands of hair (although the class finally decided that hair was not an official body part). Anna suggested fingers, and the class accepted fingers as an appropriate body part. However, as the class members stretched out their fingers on the rug and counted by tens, they realized that they would not even reach one thousand. Sam remembered how far away one million is when he wrote the reflection in figure 13.2. If one million takes your breath away, then the class clearly needed more fingers to count! All the children and

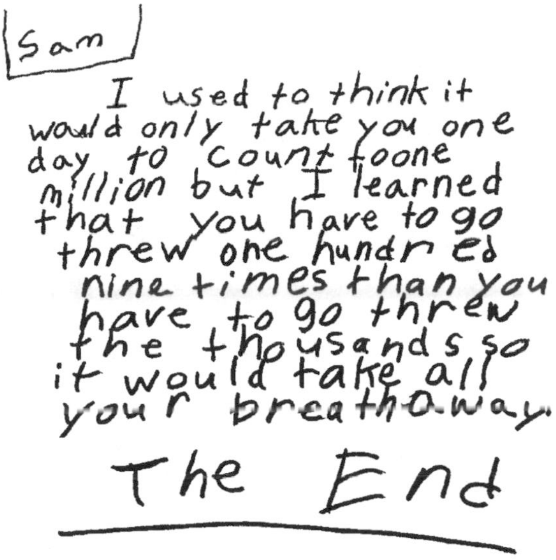

Sam

I used to think it would only take you one day to count toone million but I learned that you have to go threw one hundred nine times than you have to go threw the thousands so it would take all your breath away.

The End

Fig. 13.2. Sam's idea of one million

teachers in the second grade then lined the hallway on both sides, with their fingers outstretched. They were able to assemble 200 people, and thus 2000 fingers, and they videotaped their demonstration so other classes could appreciate their effort. Erin reflected on what she had learned from this experience, and she wrote the observation in figure 13.3. The children and teachers calculated that it would take 499 more trips with the video camera, up and down the hallway, to show one million fingers (500 × 2000). This experience helped the children develop a sense for large numbers by giving them the opportunity to devise and participate in their own unique physical representation of one million.

Students in Ms. Younce's fourth-grade class wanted to determine if their school contained one million floor tiles. They worked in pairs calculating the number of tiles in classrooms, hallways, the cafeteria, and offices. A summary

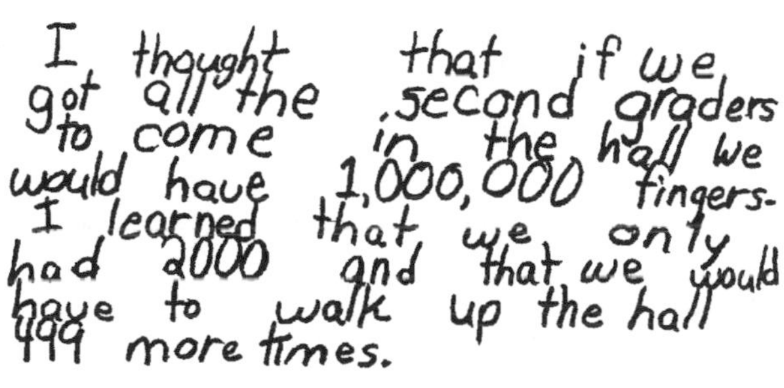

I thought that if we got all the second graders to come in the hall we would have 1,000,000 fingers. I learned that we only had 2000 and that we would have to walk up the hall 499 more times.

Fig. 13.3. Erin's reflection on one million

of their findings is included in figure 13.4. They then used a calculator to divide 1 000 000 by 53 000 to discover that they would need to build their school nineteen stories high in order to use one million floor tiles.

Mr. Stevens' fourth-grade class also wanted to relate the number one million to their own school setting. They decided to figure out how many schools would be needed to represent 1 000 000 students. They wanted to use their own school of 1100 students as the standard size. Mr. Stevens asked the students, "How many schools of this size do you think it would take to add up to one million students?" One child suggested twenty, so Mr. Stevens asked, "What do we do now that we have twenty schools with 1100 students in each school?" Some students wanted to add 1100 twenty times and others suggested multiplying. Both methods were used, and they found that 1100 × 20 = 22 000 students. They realized they were still not very close to 1 000 000, so they doubled 22 000 to get a total of 44 000 students for 40 schools. They continued to add in increments of 22 000 students. When they reached 110 000 students in 100 schools, they discussed the relationship between 100 000 and 1 000 000. Since 10 × 100 000 = 1 000 000, they multiplied 110 000 × 10 = 1 100 000 students for 1 000 schools (100 × 10). Several students insisted that this figure should be more accurate, and so they began to subtract successive groups of students and schools (fig. 13.5). They

NUMBER OF TILES FOUND

6 classrooms on fourth-grade hall	7 500 tiles
10 classrooms on first- and second-grade halls	12 500 tiles
The fourth-grade hall	2 862 tiles
The first- and second-grade hall	4 500 tiles
Teachers' lounge	500 tiles
Office	750 tiles
Principal's office	750 tiles
Assistant principal's office	750 tiles
Guidance office	750 tiles
Workroom	550 tiles
10 classrooms on third-grade hall	12 500 tiles
The third-grade hall	4 500 tiles
Cafeteria	5 000 tiles
	Approximately 53 000 tiles are in the school.

(Kindergarten rooms were added later and do not have tiles on the floors.)

Fig. 13.4. Counting floor tiles

Number of Students	Number of Schools
1 100 000	1000
− 22 000	−20
1 078 000	980
− 22 000	−20
1 056 000	960
− 22 000	−20
1 034 000	940
− 22 000	−20
1 012 000	920

Fig. 13.5. Revising the initial estimate

decided that 920 schools was a more reasonable estimate of the number of schools required to contain one million students.

Ms. Allison's third-grade class decided to take advantage of their school's mile track to demonstrate the magnitude of one million. They wanted to find out how long it would take to walk one million steps. They walked around the mile track one time and found that it took them a little more than 2000 steps to cover this distance in a span of forty minutes. Then they used calculators to determine the rest of their results:

1. 1 000 000 steps ÷ 2 000 steps per mile = 500 laps (or miles)
2. 500 laps × 40 minutes per lap = 20 000 minutes to walk 1 000 000 steps
3. 20 000 minutes ÷ 60 minutes in an hour = 333.33 hours

The class then calculated that if it were possible to keep up this pace, they would need 13.88 days of nonstop walking to walk one million steps. However, they also figured that if they walked only seven hours each school day, it would take 47.6 days. The class performed similar calculations for one billion and found it would take 47 619 school days, or 264.55 school years.

Students in Mr. Rembert's eighth-grade class wanted to represent the number one million through an area model. They were intrigued when David Schwartz described one million as equivalent to seventy pages of the stars that are shown in the book. They used a computer to create rows of asterisks on a single page. Each page had 60 rows of 81 asterisks, yielding 4860 asterisk "stars." They included several different-sized stars on each page, as well as some hot-air balloons so they could write their names on the balloons. They were not sure if they had enough space to represent one million, so they decided to represent 250 000 instead. Then they divided 250 000 by 4 860 and found they needed 51.44 pages, so they decided to use 52 pages for their display, although John kept insisting that he wanted to make the 208 (52 × 4) copies necessary to show one million.

However, when Doug checked his figure on the calculator (1 000 000 ÷ 4 860), he said he needed only 205.8 pages! They discussed how rounding accounted for the difference and then agreed that representing 250 000 (although they determined it would actually be 252 720) would be a more manageable objective. They tried to determine which classroom wall would best accommodate their six-sheet-by-nine-sheet array (minus 2 sheets) (fig. 13.6). At the conclusion of their project, they all agreed that one million was more than they had ever imagined. Even at the end, John was still wondering whether or not they ought to use more pages to represent one million, but the others convinced him that it was easy to imagine the display being four times larger, and they added, "Who really wants to tape up another 156 pages anyway?!"

Fig. 13.6

Finally, when Mr. Rembert asked Doug what he had learned from this experience, Doug referred to a topic that he was studying in science class: "In science it tells how long it takes to get to Mars. I thought about light-years. Light travels 186 000 miles a second, so I started thinking about a million. I could tell how much a light-year was 'cause these stars are 250 000." Thus, the project gave Doug a reference point for a better understanding of the magnitude of a light-year and the dimensions of our solar system.

A Final Reflection

Children's literature can play a powerful role in restoring a meaningful context to mathematical ideas. As teachers read *How Much Is a Million?* to their students, they discover that this one story can be the catalyst that helps learners raise interesting questions and pose intriguing explorations. *How Much Is a Million?* offers an open-ended invitation for children to connect their world to the world of mathematical ideas.

REFERENCES

National Council of Teachers of Mathematics. *Curriculum and Evaluation Standards for School Mathematics.* Reston, Va.: The Council, 1989.

Schwartz, David. *How Much Is a Million?* New York: Lothrop, Lee, & Shepard Books, 1985.

Whitin, David J., and Sandra Wilde. *Read Any Good Math Lately?* Portsmouth, N.H.: Heinemann Educational Books, 1992.

14

Connecting Reasoning and Writing in Student "How to" Manuals

Neal F. Grandgenett
John W. Hill
Carol V. Lloyd

ANYONE who has ever purchased a do-it-yourself kit for products such as a birdhouse or swing set and then spent hours putting it together will tell you that following the instructions for assembly in the included manual can be a real exercise in mathematical reasoning. However, the person who really exercised reasoning skills was the person who developed the manual in the first place. We found that having students create how-to manuals is a nice way to blend reasoning and writing in the elementary school classroom while involving them in a stimulating, real-life mathematical task. We successfully incorporated this idea into a five-week summer clinic for elementary school at-risk students and students with learning disabilities.

The how-to manuals that we had each student create were small booklets that listed and illustrated the steps to building a particular product, such as a paper box kite, similar to the structure of real how-to manuals. The main purpose of the exercise was to help the students develop a type of mathematical reasoning called *procedural reasoning.* This reasoning skill involves breaking a process down into smaller, distinct steps; it is consistent with the National Council of Teachers of Mathematics (NCTM) reasoning standards (NCTM 1989) and is a common reasoning skill practiced in such activities as creating a computer program, writing a recipe, or giving someone directions. As students planned, built, and documented their constructions with their manuals, they were also involved in such mathematical activities as measuring with a ruler, making arithmetic calculations, and drawing three-dimensional diagrams.

Since improved student writing was also an important goal for our summer clinic, we carefully organized the learning activities in a process-writing approach. This approach involves students in brainstorming ideas, writing down their ideas, conferring with other students, revising their work, and, finally, publishing their work (Graves 1985; Graves and Hansen 1984). Since the how-to manuals were structured written work expressing the students' reasoning steps, the process-writing approach was a useful way both to organize instruction and to ensure that students were assisted in the writing process. The instructional activities were then broken down into four process-writing phases.

INSTRUCTIONAL ACTIVITIES

Phase 1: Construction

To begin the construction phase, we first led the class in a planning discussion of the dimensions and characteristics of a product model, in this instance a model box kite. Students were asked such questions as "How long is the kite?" and "What is the area of its paper surface?" The students then took turns finding the needed measurements, doing the calculations, and sharing the results with the class. We also asked thought-provoking, science-related questions such as "What makes a real box kite fly?" "What are the differences between your model kite and a real kite?" and "Could you attach the string to anywhere on a real kite?"

After this initial planning discussion, we asked the students to make a box kite similar to a teacher-made model. Since we had students ranging from first to seventh grade, the products we used varied in sophistication from student to student, but all students worked on the same type of product each week, such as a box kite, a box house, and a box space station. Students had access to construction materials that included poster board, glue, plastic straws, and tape, as well as construction tools such as rulers, scissors, and pencils. They also had access to the school library, where they could seek additional information about their product as desired. For instance, several students found it necessary to research whether their box kite needed a tail or not.

All the products used student-created boxes that were made out of 100-square-inch pieces of poster board, which required students first to measure and cut 10-inch-by-10-inch squares from a larger sheet. Additional mathematical concepts were introduced as the squares were folded into boxes. For instance, students investigated the concept of volume by trying to decide which dimensions would produce a box with the largest volume and then tested possibilities with teacher-supplied jelly beans. Students then used their simple boxes to create the more elaborate models of real products, such as a box-kite model, composed of two boxes separated by four sticks (we used soda straws).

Phase 2: Writing

After the students constructed their products, they were asked to record the steps they used to build them. Students sat near their "creations" and remembered the steps they used in constructing them. With paper and pencil, they wrote down their steps in the construction process, with the teacher reminding them periodically that they were writing out the directions so that some other student could also build their product. We also encouraged them to pick up and reexamine their products, or even take them partially apart, to help them remember the steps they had used in the construction process. Many students requested to type out their steps on the word processor so that their directions would be easier for other students to read. Students did a surprising amount of careful writing during this phase, as shown by a fourth-grade student's written steps in figure 14.1.

HOW TO MAKE A BOX KITE

1. Get a big piece of paper.
2. Then draw a big square.
3. Get a ruler.
4. Use the ruler to draw eather a 1-inch 2-inch or 3 inch going down
5. Whatever you pick cut the side of the square to the line.
6. Then you fold up all the sides and tape them.
7. Get 4 straws and tape them in the corners.
8. And make another box and tape the four straws to the corners of the other box.

YOU HAVE A BOX KITE!!!!!!!

Fig. 14.1. A fourth-grade student's written steps that describe how to build a box kite

Phase 3: Revision

Next, the students shared their written steps with at least one other student to help refine their manuals. This peer conferencing allowed students to ask each other such questions as "How long a line?" or "Where did you fold the paper?" The students were very supportive of one another during this activity, and they seemed genuinely curious about what other students had listed for construction steps. After consulting, the students worked on revising their written steps.

Phase 4: Publication

Finally, the students created diagrams to help illustrate their how-to manuals. Many of the students cut apart their written steps and tried to put a diagram with each step. This resulted in steps that resembled the real-life how-to manuals found in prefabricated kits such as model cars or rockets. A third-grade student's example is shown in figure 14.2.

Fig. 14.2. A third-grade student's step mixing text and diagrams

A title page was added, which usually involved a title such as "How to Build a Box Kite" with the student's name listed as the author. Many of the manuals were quite elaborate and used multiple pages of written steps and diagrams. Some students even included diagrams of construction materials and of objects in three dimensions, such as the fourth grader's work shown in figure 14.3.

Fig. 14.3. A fourth-grade student's diagram to illustrate the needed construction materials

ASSESSMENT

Assessment was an integral part of the activity. For instance, during the construction phase it was easy to tell which students needed additional instruction in some of the more fundamental skills, such as how to use a ruler. Students themselves also contributed to the ongoing assessment by helping to review and critique the work of other students. The most rewarding assessment, however, came at the end of the four phases when students were encouraged to have other students build a product using their manuals. They found this challenge to be a true test of the accuracy and utility of their work.

THE POWER OF THE "CONSTRUCTION CONNECTION"

In summary, we found that student how-to manuals can produce an enjoyable and natural connection between mathematics and writing in the elementary school classroom. Blending reasoning with writing helps develop both these important skills, and it furnishes the opportunity to apply basic concepts of measurement, arithmetic, and geometry. Science can also be represented, since teachers have the opportunity to ask students thought-provoking questions such as "Why does a box kite fly?" We even found that some important concepts of art, such as perspective drawing, were also represented as students included two- and three-dimensional diagrams in their manuals. We were proud that during the different phases of the summer project our students became, in turn, architects, writers, scientists, and artists as they developed their how-to manuals. Yet the real power of our activities became apparent when we watched these students excitedly share their how-to manuals with one another and with their parents. It seemed that our "construction connection" had helped some youngsters who had become used to failing to gain some real experiences with success in the mathematics classroom.

REFERENCES

Graves, Don. "All Children Can Write." *Learning Disabilities Focus* 1 (1) (1985): 36–43.

Graves, Don, and Jane Hansen. "The Author's Chair." In *Composing and Comprehending,* edited by Julie M. Jensen. Urbana, Ill.: National Council of Research in English, 1984.

National Council of Teachers of Mathematics. *Curriculum and Evaluation Standards for School Mathematics.* Reston, Va.: The Council, 1989.

15

Connecting Mathematics and Physical Education through Spatial Awareness

Diana V. Lambdin
Dolly Lambdin

THE development of spatial awareness is an important goal both for mathematics teaching and for physical education. This paper looks at spatial awareness in the physical education curriculum and at research on how spatial concepts are learned, and it gives suggestions for activities to integrate mathematics teaching with physical education.

BEHIND THE SCENES IN PHYSICAL EDUCATION

There are two very different types of curricula in elementary school physical education. The traditional organizing framework is through sports, games, dance, and gymnastics. However, in recent years many physical educators have adopted a movement-based curriculum, especially in the primary grades. Problem solving and guided-discovery teaching are common in this curriculum where movement itself is studied using the interaction of types of movement (locomotor, nonlocomotor, and manipulative) with modifying concepts grouped under the categories of spatial awareness and relationships (see table 15.1).

The goal of movement-based physical education is to help children become comfortable applying basic movement skills to a wide variety of activities. Students learn such skills as how to move creatively without bumping into others and how to use all the space available. Activities commonly require students to experiment with using different modes and speeds of travel, following a variety of pathways, and focusing on different body shapes.

Partner and group routines often emphasize spatial relationships among participants, requiring either matching, mirroring, or contrasting body

TABLE 15.1
Spatial Concepts Used in Movement Education

Spatial Awareness	
location:	self-space and general space
directions:	up/down, forward/backward, right/left, clockwise/counterclockwise
levels:	low/middle/high
pathways:	straight/curved/zigzag
extensions:	large/small, far/near
Spatial Relationships	
of body parts:	round/narrow/wide/twisted/ symmetrical/asymmetrical
with objects:	over/under, on/off, near/far
with other people:	matching, mirroring, following

Table adapted from Graham, Holt-Hale, and Parker (1993)

shapes. Initial routines usually deal simply with the body and with space. As students develop skill in using spatial concepts, large or small equipment is added as well as additional people to form cooperative partner or group activities. Once the concepts of levels, pathways, shape, symmetry and asymmetry, and relationships are understood, they are used in activities and lessons where the emphasis is on other skills, such as throwing and catching, jumping and landing, or expressive movement.

Beyond the spatial awareness developed by movement-based physical education curricula, traditional physical education activities in gymnastics, dance, and team sports also require considerable spatial sense and can furnish interesting examples for mathematics classes.

In gymnastics, body shape is extremely important, with symmetry required in some moves and asymmetry in others. Tucked shapes must be tightly curled, piked shapes require straight legs, and layout positions demonstrate a straight body (see figure 15.1). In floor exercise, the diagonal of the mat is

Fig. 15.1. Adapted from Kreighbaum and Barthels (1990)

used for the tumbling runs because speed is needed for these difficult stunts, and the diagonal provides the longest run for developing speed.

Folk dances require spatial sense both in formations (squares, circles, double circles, etc.) and in the laterality and directionality required for performance. Students must learn to move to their own right while seeing those across the circle from them moving in the opposite direction. Often, individuals will turn clockwise while the entire circle is moving counterclockwise. Being able to picture groupings and pathways is crucial to successful movement in these activities.

In team sports, knowing the correct trajectory or pathway of the ball is important to the execution of many skills. In basketball, the chest pass, bounce pass, and lob pass all represent different pathways of the ball for specific situations; to maximize distance, the ball must be projected at an angle close to 45 degrees. Players must learn to judge the pathway needed for their purpose. Whereas a soccer goalie might choose a 45-degree angle of projection to punt the ball as far down the field as possible, a football punter might choose a steeper angle (for a shorter distance) to allow the defensive players to reach the receiver as soon as the ball is caught.

Personal pathways also play a large part in team sports. The quickest way to move downfield from one point to another is to use a straight pathway. Fakes require a zigzag pathway to lose the opponent, and curved pathways are often used to create spaces for future movement. Football receivers must know specific pathways so that the quarterback can pass the ball to the right spot, releasing the ball before the receiver is there. For example, a *down-and-out* requires a run straight ahead, then a 90-degree turn toward the sideline. A *slant* has the receiver running toward the corner of the field. A *down, out, and down* requires two 90-degree changes of direction.

RESEARCH ON HOW SPATIAL CONCEPTS ARE LEARNED

Both physical education and classroom teachers can benefit from familiarity with research on how pupils learn spatial concepts. In a very teacher-friendly article, Del Grande (1987) briefly defined such abilities as eye-motor coordination, figure-ground perception, perceptual constancy, and position-in-space perception and then described activities appropriate for developing, in the mathematics classroom, these abilities that are often the focus of attention in physical education classes as well.

Also useful in planning spatial-awareness lessons is familiarity with the work of Pierre van Hiele and Dina van Hiele-Geldof, Dutch researchers who proposed a five-level hierarchy in the learning of geometric concepts. At the lowest van Hiele level, Level 0, students think holistically about shapes, primarily through identifying, naming, and comparing them. When students move to Level 1, they begin to analyze figures in terms of their components (e.g., a cube is composed of faces, edges, and vertices) and in terms of relationships among those components (e.g., opposite faces on a cube are parallel). At higher van Hiele levels, students prove previously discovered properties first through informal logic and later through formal deduction. However, very few elementary school pupils exhibit thinking beyond Level 1. (See Crowley [1987] for a concise explication of the van Hieles' work.)

INTEGRATION OF THE CONCEPTS

When teachers plan together, physical education and mathematics instruction can complement each other. Jensen and Spector (1987) give a number of ideas for activities relating movement and spatial awareness. For example, they suggest having pupils imagine that they are in a cubical space capsule: the children move around "within" the capsule, "feeling" the sides, "touching" and counting the faces, edges, and vertices, and generally experiencing movement constrained by the boundaries of the imaginary cube. A few more integrated activities follow.

Devising and Performing Gymnastic Routines

In a unit planned jointly by their classroom teacher and their physical education teacher, pupils might (1) learn a number of dance or gymnastics moves in physical education; (2) learn about various types of symmetry (e.g., line symmetry, point symmetry) and about certain geometric transformations (e.g., slide, flip, turn) in mathematics class; (3) work in small groups in the mathematics classroom to diagram on paper a gymnastic routine using physical moves involving symmetry and transformational concepts (following each other, mirroring each other, twisting, turning, flipping, etc.); (4) practice and refine their gymnastics routines in the physical education class; and, finally, (5) perform their finished routine at a school assembly or parents' night.

Using a Movement-Sized Grid

Teachers might collaboratively plan an activity to engage students in using geoboards to make specified shapes, copying these shapes onto dot paper, and then following their dot-paper "maps" to perform spatial awareness activities on a movement-sized grid on the playground or in the gymnasium. (Vinyl polyspots or cones, available from physical education supply catalogs, can be thrown down to form a grid wherever needed. On wood, tile, or blacktop, colored tape or shoe polish can be used. Velcro dots are useful on carpeted floors.) In addition to basic traveling tasks, speed or locomotor pathways can be prescribed on specific sides of a geometric shape (e.g., "Follow a rectangular pathway traveling backward along the short sides and forward along the long sides" or "Use a right triangle for a pathway; walk along the legs and skip along the hypotenuse"). More advanced pathways from shapes developed in mathematics class can be used in the development of routines in gymnastics or expressive dance.

Investigating Distance around a Track

Another collaborative effort to develop spatial awareness might involve students in measuring the distance around a running track in each of its lanes and then determining where the starting blocks should be in each lane for various lengths of races. Although pupils, when running a race, may naturally gravitate toward the center of the track, they often are not aware of why it is advantageous to run in the innermost lane. A related activity might investigate why, in a folk dance where lines of dancers form spokes of a wheel to move around a center point, dancers near the outside must dance much faster than those near the center, who often merely march in place.

CONCLUSION

Activities that help students develop spatial awareness furnish a natural connection between mathematics and physical education. Why not develop a more collaborative atmosphere in your school by sharing ideas about spatial awareness with your physical education specialist or with other classroom teachers?

REFERENCES

Crowley, Mary L. "The van Hiele Model of the Development of Geometric Thought." In *Learning and Teaching Geometry, K–12,* 1987 Yearbook of the National Council of Teachers of Mathematics, edited by Mary Montgomery Lindquist, pp. 1–16. Reston, Va.: The Council, 1987.

Del Grande, John J. "Spatial Perception and Primary Geometry." In *Learning and Teaching Geometry, K–12,* 1987 Yearbook of the National Council of Teachers of Mathematics, edited by Mary Montgomery Lindquist, pp. 126–35. Reston, Va.: The Council, 1987.

Graham, George, Shirley Holt-Hale, and Melissa Parker. *Children Moving: A Reflective Approach to Teaching Physical Education.* 3rd ed. Mountain View, Calif.: Mayfield Publishing, 1993.

Jensen, Rosalie, and Deborah C. Spector. "Geometry Links the Two Spheres." In *Geometry for Grades K–6: Readings from the "Arithmetic Teacher,"* edited by Jane M. Hill, pp. 170–73. Reston, Va.: National Council of Teachers of Mathematics, 1987.

Kreighbaum, Ellen, and Katharine Barthels. *Biomechanics: A Qualitative Approach for Studying Human Movement.* 3rd ed. New York: Macmillan Publishing Co., 1990.

PART 4

Connections across the Middle School Curriculum

16

Seeing and Thinking Mathematically in the Middle School

Glenn M. Kleiman

A NEW curriculum for middle schools called Seeing and Thinking Mathematically builds on the following view of mathematics as central to the human experience (Kleiman 1991, pp. 48–51).

> Mathematics is, and has always been, part of the essence of being human. To be human is to seek to understand. Mathematics, along with science, has made possible dramatic advances in our understanding of the physical universe. To be human is to explore. Throughout history, mathematics has been essential for exploration, from navigating by the stars to travel into space. To be human is to participate in a society. Societies require mathematics to keep records, allocate resources, and make decisions. To be human is to build, and mathematics is essential for the design and construction of everything from tents to temples to skyscrapers. To be human is to look to the future. Mathematics enables us to analyze what has been, predict what might be, and evaluate our options. To be human is to play, and mathematics is part of our games and our sports. To be human is to think, to create, and to communicate. Mathematics provides a vehicle for thinking, a medium for creating, and a language for communicating. Indeed, to be human is to develop mathematics. Mathematics has been developed in every culture for the purposes of counting, locating, measuring, designing, playing, and explaining (D'Ambrosio, 1985; ... Gilmer, 1990). Without mathematics, as without language, the nature of humanity and human society would be fundamentally different.

The author thanks the staff, consultants, teachers, and students involved in the Seeing and Thinking Mathematically project, all of whom have contributed to the ideas in this paper. The author also thanks Leigh Peake for her suggestions and editorial assistance in preparing this paper. The preparation of this paper and the development of the curriculum described in it have been supported by grant no. MDR-905-4677 from the National Science Foundation. The views expressed in this paper and in the curriculum do not necessarily reflect those of the National Science Foundation.

The new curriculum is organized around thematic units in which students engage in designing, building, planning, creating, playing, analyzing, deciding, experimenting, and communicating, that is, using mathematics in the ways it has been used in all societies throughout history. These thematic challenges lead to investigations that enable students to develop their mathematical understandings and techniques. Examples of some thematic challenges and connected mathematical investigations are given in figure 16.1.

Sample Thematic Challenges	Representative Mathematical Investigations
Determine how to make a building more accessible to handicapped students.	Explore the relationship between the angles and sides of a right triangle.
Determine which cereals should be purchased for a school breakfast.	Determine the cost for each serving and for other quantities.
Build a house from geometric shapes. Create building plans, using pictures and written instructions, so that others will be able to build the same house.	Represent three-dimensional shapes in two dimensions with nets, orthogonal drawings, perspective drawings, and verbal descriptions.
Determine whether a game of chance is fair for all the participants.	Find all possible outcomes with tree diagrams and outcome grids.
Identify patterns in the number words of a language that make it easier to learn to count in that language.	Explore number patterns and the use of addition, subtraction, and multiplication in the composition of numbers.
Analyze changes in the population of a town to plan whether new schools will be needed in the future.	Explore the differential results of linear and geometric growth.
Create a familiar object that you have scaled to the size it would be in Brobdingnag, the land of giants that Gulliver visited.	Determine the effects of rescaling on linear, area, and volume measurements.

Fig. 16.1. Sample thematic challenges and mathematical investigations

When faced with thematic challenges such as those listed in figure 16.1, students call on all their available mathematical knowledge and problem-solving expertise. They do not ask, "Are we doing computation, geometry, or statistics today?" but rather, "What do we need to figure out? What do we know that might help? And how might we go about it?" For any given problem, students may use very different approaches. For example, some may do many calculations to create a table, whereas others may use a graph to estimate and still others may use algebraic equations. As the students share and compare approaches, relationships among the approaches come to light and students can add to their repertoires of mathematical techniques.

Since students learn mathematics in the context of real uses of mathematical thinking, the concepts and processes they master give them capabilities they can use outside of school. Although their daily lives may not demand that they undertake tasks like proving the Pythagorean theorem, students do need to make decisions, share things fairly, plan, build, play, communicate, and understand the world around them. In some of these activities, such as building, students may find the Pythagorean theorem to be very useful.

AN EXEMPLARY UNIT: DESIGNING SPACES FOR PEOPLE

In the unit "Designing Spaces for People," students explore the uses of mathematics in designing and building houses. The unit begins with students' considering why houses in different places have different shapes (e.g., an igloo versus an A-frame). Students then design, build, measure, represent, and describe their own structures. Through the investigations in this unit, students learn to—

- build three-dimensional structures from two-dimensional shapes;
- represent three-dimensional structures in two dimensions with floor plans, nets, and verbal descriptions;
- determine the areas of various shapes to find the amount of material used to build floors, walls, and roofs;
- determine the cost of a building from the unit costs of materials.

A House for a Hot, Rainy Climate

In one activity, students build a model house for a tropical (hot and rainy) climate using a set of geometric shapes, including squares, equilateral triangles, trapezoids, rhombuses, pentagons, hexagons, and one additional shape that the students select. This activity is designed to give students hands-on experience with a challenge similar to that faced by people throughout history: designing a shelter for a given climate using limited available materials. Students design houses with sloped roofs, domed roofs, overhangs, ventilation systems, and stilts to raise the house above floods. Figure 16.2 shows the variety of structures designed by students in one San Francisco classroom.

Mathematical Connections

Building a house for a specific climate from geometric shapes leads to a number of mathematical questions involving geometry, measurement, and computation. Some examples are the following:

- The roof in a rainy climate cannot have holes in it. What pieces can be fit together to make a solid roof?

Fig. 16.2

- How can we make the roof slope so that rain runs off it? How much should it slope? How can we describe the amount of slope?
- How can we design a roof that will be the right size and shape to fit on the base of the house?
- If the amount of material in a square unit costs $100 for floors and walls and $200 for roofs, how much does the material for the entire building cost?

Connections to Other Subject Areas

The themes of architecture and construction lead to connections to geography, social studies, history, art, and science. Interdisciplinary questions that flow naturally from the unit's activities include the following:

- How do the climate and the available building materials differ around the world? How do the houses in different locations reflect these variations?
- How do housing needs differ in different societies (e.g., nomadic versus farming societies)?
- How do houses from various periods in history compare to modern houses? What role have new technologies and materials played in the changes?
- How do architects design buildings? How does a construction company build a structure from the architect's plans?

- How do building materials differ in strength, flexibility, weight, and durability? How do these properties determine which materials can be used for different purposes (e.g., could a skyscraper be built of wood)?

Connections to the Local Community

Buildings are found in every community. Students can investigate the architecture of local buildings, the zoning laws that determine what can be built where, and the building codes. They can build models of local buildings and consider how the design of buildings in their community could be improved. For example, they might make a plan for improving the handicap access to public buildings in the community.

Multicultural Connections

The information about buildings, materials, and climates that people from other countries can bring into the classroom enrich the activities for the entire class. For example, a student from Japan taught one class about rice paper and how it is used in Japanese structures. Students from Cambodia contributed the idea of using stilts to raise the buildings so they would not be damaged by floods. (See Zaslavsky [1989] and Whitin [1993] for more activities exploring the mathematics of buildings from around the world.)

CONCLUSIONS

Seeing and thinking mathematically is universal—an essential aspect of all cultures for all of history. Using mathematics for culturally significant purposes leads to a curriculum that is rich in mathematical connections, interdisciplinary connections, community connections, and multicultural connections. In this type of culturally open curriculum, students from every country and culture can contribute examples of games, buildings, designs, stories, maps, languages, and computational techniques that can enrich the thematic challenges for all the students in the class (see Ascher [1991] for examples from the field of ethnomathematics; see also the article by Shirley in this yearbook). This paper presents just a few examples of this approach that can make the most important connection of all: that of mathematics to the lives of students.

REFERENCES

Ascher, Marcia. *Ethnomathematics: A Multicultural View of Mathematical Ideas.* Pacific Grove, Calif.: Brooks/Cole Publishing Co., 1991.

D'Ambrosio, Ubiratan. *Socio-Cultural Bases for Mathematics Education.* Campinas, Brazil: UNICAMP, 1985.

Gilmer, Gloria. "Ethnomathematics." Paper presented at the annual meeting of the National Council of Teachers of Mathematics, Salt Lake City, Utah, April 1990.

Kleiman, Glenn M. "Mathematics across the Curriculum." *Educational Leadership* 49 (October 1991): 48–51.

Whitin, David J. "Looking at the World from a Mathematical Perspective." *Arithmetic Teacher* 40 (April 1993): 438–41.

Zaslavsky, Claudia. "People Who Live in Round Houses." *Arithmetic Teacher* 37 (September 1989): 18–21.

17

Projects in the Middle School Mathematics Curriculum

Stephen Krulik
Jesse Rudnick

It has long been common practice in social studies, English, science, and other school subjects for teachers to create and assign projects to groups of students—cooperative learning activities that extend over several consecutive class periods or even over several weeks of class. Projects have been less common in mathematics classes, where students may spend considerable time learning and practicing specific algorithmic skills. Yet there are projects that not only use and strengthen algorithmic skills but also develop process skills and connect mathematics to other areas of human endeavor and to other school subjects.

This article outlines two projects that can be used at the middle school level to connect the mathematics specified in the curriculum to other branches of mathematics, to other school subjects, and most important, to the students' daily lives. Students find the projects interesting, enjoyable, and helpful in developing a "feel" for the uses of mathematics outside class. Many such projects can be undertaken jointly among classes in other subject areas, including economics, science, language arts, social studies, mathematics, and more.

Why Projects?

According to *Webster's New World Dictionary* (2nd college ed.), a project is "a proposal of something to be done, a scheme." In mathematics education, a project may assume a much broader meaning, one that includes not only deciding what must be done but also doing it, presenting the data, and assessing the findings or results. The project should not be completely defined by the teacher; rather, it should be a group effort with the teacher acting as a guide. School projects parallel real life when students are asked to recognize the existence of a problem, define the problem, set the parameters for what is to be done, decide what to do and how to do it, assign specific tasks, carry out the work, and present the results to a larger group in a clear and succinct format.

Projects offer an excellent opportunity for students to become involved in cooperative learning, since projects are generally assigned to groups of students. This helps students sharpen their communication skills as they talk with other members of their own group and communicate their findings and results to the entire class in both oral and written form. Working together on cooperative projects also helps students develop the ability to work with others, a necessary life skill. Projects also furnish opportunities for creative individual thinking as well as for group discussion and thought. Students must answer questions such as these: What do we need to do to resolve the problem? Where can we obtain the needed data? What comes next? How can we apportion the tasks? What role will we each play in the research? How can we best keep track of our results? What is the best way to report our results?

PROJECT GULLIVER

After your fifth-grade students have read *Gulliver's Travels* (several abridged editions are available for younger students), ask them to describe your classroom as if it were in Lilliput or in Brobdingnag. Tell them that an average fifth-grade student in Lilliput is seven inches tall; in Brobdingnag, an average fifth grader is 18 feet tall. Students should include a description, drawings and dimensions for the classroom, a student's desk, a teacher's desk, a textbook, a notebook, a pencil, and so on.

Teacher's Notes

The students' first task is to determine the ratio of the average height of a Lilliputian (or Brobdingnagian) to the average height of an American fifth grader. There are at least three ways to do this:

1. Find the average height of the entire class.
2. Find the average height of the boys in the class and of the girls in the class separately. Then average the results.
3. Find the average height of the students in each cooperative group and use these data to project the average height of the fifth-grade class.

Compare and discuss the results of these efforts. (Note that method 2 is correct only if there are the same number of boys as girls in the class.) National statistics show that the average height of a fifth-grade student is about 55 inches.

After your students have worked in groups to figure out the average heights of the fifth-grade Lilliputians and Brobdingnagians, have them determine the ratios of the average height of American fifth graders to the average heights of their Lilliputian and Brobdingnagian fifth-grade counterparts. They should arrive at the following ratios:

$$\frac{H_L}{H} = \frac{7}{55} = 0.127,$$

which is a ratio of approximately 1:8, and

$$\frac{H_B}{H} = \frac{18'}{55''} = \frac{18 \cdot 12}{55} = \frac{216}{55} = 3.93,$$

which is a ratio of approximately 4:1.

Once these ratios have been determined, use the students' knowledge of ratio and proportion to scale items in the three different-sized classrooms. For example, a teacher's desk that measures 42 inches long, 31 inches wide, and 30 inches high in our classroom would measure 5.25 inches long, 3.875 inches wide, and 3.75 inches high in Lilliput, if properly scaled. In Brobdingnag, however, this same desk would measure 168 inches long (14 feet), 124 inches wide (10 feet 4 inches) and 120 inches high (10 feet). A pencil that is 7.5 inches long and has a diameter of 0.25 inch in our classroom would measure a little less than 1 inch in length (0.938 inch) and would have a diameter of 0.03 inch in Lilliput; in Brobdingnag, the pencil would be 2.5 feet long and have a 1-inch diameter.

The discussion above is limited to linear measures only; as an extension, students might consider relative weights in the three different populations. Keep in mind, however, that weight varies with volume and is, therefore, a cubic relationship. Thus, if an American child is eight times the height of a Lilliputian, the American would be 512 times as heavy.

Connections

Language Arts. Assume you are visiting a family in either Lilliput or Brobdingnag. Write a story about a day during your visit.

Health Science and Nutrition. You have two visitors, a fifth grader from Lilliput and a fifth grader from Brobdingnag. They are staying with you and will accompany you to school for the day, and you are packing lunch. Describe the three lunches. (Note that the food required is, again, related to volume and is therefore not a linear relationship.)

Art. Build a model of a textbook, a pencil, or another object found in a Lilliputian or Brobdingnagian classroom.

Physical Education. You will have a track-and-field meet in which the teams will come from your class, Lilliput, and Brobdingnag. Create the series of events that you will use and tell how you can make them fair for all competitors. For example, if American fifth graders run a 50-yard dash, Lilliputian students should run the 19-foot dash, and the Brobdingnagians should run the 600-foot dash at the same time.

PROJECT TRAFFIC LIGHT

Traffic lights should remain red or green for a timed interval depending on the number of cars that proceed in each direction. Are the traffic lights nearest your school accurately timed?

Teacher's Notes

Traffic engineers spend a considerable amount of time, energy, and money to regulate traffic flow. Correctly timed traffic lights are fundamental to the process. This project seeks to determine if the traffic lights in the vicinity of the school are timed correctly.

Divide the class into groups of five students. Two students in each group should time the periods that the light stays green; two other students in each group should count the number of cars that proceed through the intersection in both directions. A fifth student records the results. Be certain that the experiment is performed at several different times during the school day. Have groups combine their results and use ratios to determine the relative traffic flow at the intersection compared to the relative timing of the traffic lights. Are the ratios equal? Do the ratios remain constant all day?

Connections

Science. Automobiles are a major source of air pollution today. The students can investigate pollution controls being enforced by state or local traffic authorities. Consider whether or not excessively long red lights are forcing traffic to back up and thus add to the air pollution near the school.

Language Arts. If the ratio of traffic control to traffic flow is not compatible, the students should write a letter to the traffic commissioner in the city, informing him or her of their results.

Social Studies/Civics. Have the students visit city hall and speak with the traffic engineers to determine what surveys were conducted before the timing of the lights was set. Students might present their data to the engineers.

The traffic-light problem can also be extended to the timing of traffic lights on one-way streets, main arteries, and rush-hour designs.

Projects such as these furnish an opportunity for collaborative learning, making connections, and relating classroom mathematics to the child's world outside school, and they help teachers attune their mathematics teaching to achieve the goals suggested in the National Council of Teachers of Mathematics *Curriculum and Evaluation Standards for School Mathematics* (1989).

REFERENCE

National Council of Teachers of Mathematics. *Curriculum and Evaluation Standards for School Mathematics.* Reston, Va.: The Council, 1989.

18

Carpet Laying: An Illustration of Everyday Mathematics

Joanna O. Masingila

ALL students bring to school mathematical knowledge gained from everyday situations they have experienced. However, such knowledge is often hidden and unused in school as students learn to perform the mathematical procedures that teachers demonstrate and evaluate. According to Lester (1989, p. 34):

> The informal, sensible methods children have learned outside of school are ignored or discouraged, and little oral arithmetic or substantive discussion is related to the meaning of the formal procedures that are taught. Thus, it is not surprising that students spend very little time using their intuition or making sense of what they do in school mathematics; they rarely are expected or given the opportunity to do so.

Students need in-school mathematical experiences to build on and formalize the mathematical knowledge gained in out-of-school situations. An important part of a mathematical experience in school is the guidance and structure that can be offered by teachers who help students make connections among mathematical ideas encountered in everyday situations and in school situations. The *Curriculum and Evaluation Standards for School Mathematics* (National Council of Teachers of Mathematics [NCTM] 1989) emphasizes that connections must be made between doing mathematics in school and doing mathematics in the everyday world. The *Professional Standards for Teaching Mathematics* (NCTM 1991) expands on this theme of making connections by suggesting that teachers should build on the knowledge students bring to the mathematics classroom and should create environments in which students learn mathematics in meaningful contexts, contexts connected to students' out-of-school experiences.

MEASUREMENT IN A CARPET-LAYING CONTEXT

Measurement is an important part of the school mathematics curriculum. In fact, the *Curriculum and Evaluation Standards* states that "measurement is of central importance to the curriculum because of its power to help children see that mathematics is useful in everyday life and to help them develop many mathematical concepts and skills" (NCTM 1989, p. 51). Although measurement experiences in school are often limited to memorizing formulas (e.g., $P = 2 \cdot (l + w)$ and $A = l \cdot w$) and learning measurement skills such as how to use a ruler, measurement in everyday situations is often composed primarily of concepts and processes central to measuring, such as area expressed as square units, estimation, spatial visualization, minimizing error, and efficiency.

Measurement is a central concept, and measuring is a central process in carpet-laying work; problems from this context can be used to engage students in exploring measurement ideas. In addition, situations in the carpet-laying context lend themselves easily to problem solving because of the numerous constraints involved (Masingila 1992). The two problems discussed below illustrate how the carpet-laying context can be used to help students connect measurement ideas in school with measurement ideas and their use in out-of-school situations. Although students may not have prior firsthand experience with estimating and installing carpet, most will have some general knowledge about carpet.

> **Problem 1:** Carpet the room shown in figure 18.1 with the given piece of carpet. The square in the middle of the room is a post that extends from the floor to the ceiling. Decide how to install the carpet so that the number of seams is minimized and use scissors to cut the necessary seams.

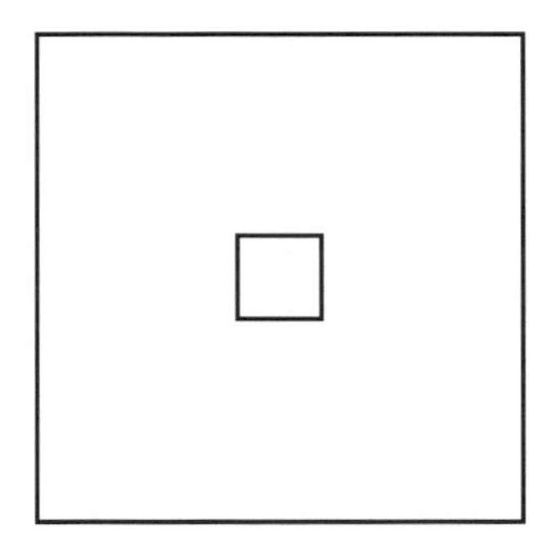

Fig. 18.1. Room with a post

> **Problem 2:** Find the measurements of the room shown in figure 18.2. The diagram is drawn on a scale of 1/4 inch = 1 foot. Find the most cost-efficient way to carpet this room given the following constraints: (*a*) carpet pieces are 12 feet wide, (*b*) the nap of different carpet pieces must all run in the same direction, and (*c*) seams should be placed out of normal traffic patterns whenever possible. Note that two inches must be added to the width and length measurements to allow for trimming,

and three inches must be added to the measurement of a dimension if the carpet must extend into a doorway. How much carpet will be needed and how much will it cost if the carpet is $23.95 a square yard?

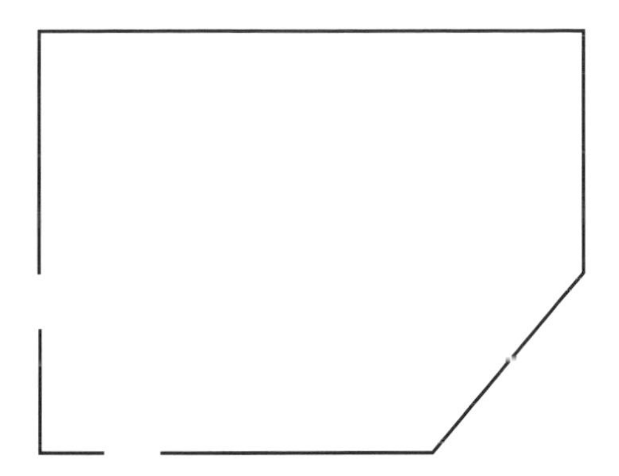

Fig. 18.2. Room to carpet

These two problems can involve students in doing mathematics in a way that is similar to how carpet layers do mathematics—measuring, making decisions, testing possibilities, and estimating in a natural way as the situation dictates. Both of the problems contain real-life constraints like those present in any carpeting job. Problem 1 engages students in thinking about minimizing seams while installing carpet around a post. To solve problem 2, students must weigh cost efficiency against seam placement. At the same time, they must make sure that the nap is running in the same direction for all the carpet pieces. These problems also provide good situations for cooperative problem solving and fruitful discussions about optimal solutions.

SOLUTIONS

Students are likely to generate a variety of solutions to the two problems above. Some solutions will be better than others because they account for the constraints in a more efficient manner. The following are solutions generated by pairs of ninth-grade general mathematics students and by a carpet installer and estimator.

Problem 1

Students' Solutions. Shown in figure 18.3 are solutions of three different pairs of students; the lines indicate seams. The students worked with a floor plan drawn on paper and a piece of paper "carpet." A three-dimensional model of the room with a rectangular prism post sticking up could provide an even better illustration of the constraint posed by the post. In the first two solutions, the students decided that the carpet piece needed to be cut into two pieces before fitting it around the post. In the third solution, the students decided to cut out a section of the carpet and replace part of the section after fitting the carpet around the post.

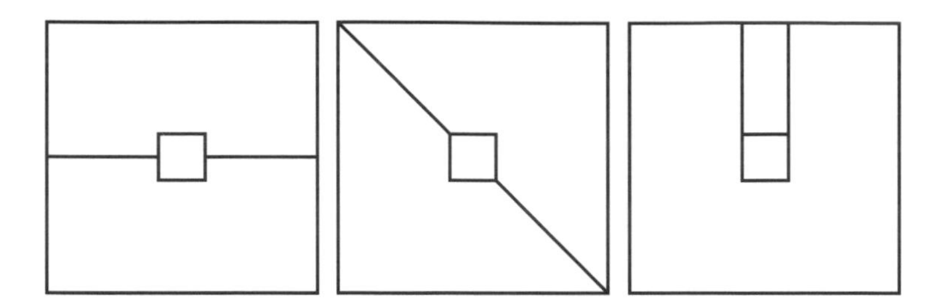

Fig. 18.3. Students' solutions

Installer's Solution. The installer who faced this problem solved it in the following manner. He took the rolled-up carpet to a wall, laid the carpet out up to the post, and then folded the rest back. He made a cut from one corner of the post half the distance of one face of the post. Next he made a perpendicular cut from the center of the post all the way to the edge of the carpet. The installer then cut the carpet along the sides of the post so that the carpet was laid with only one seam from the post to the wall on one side of the post (fig. 18.4).

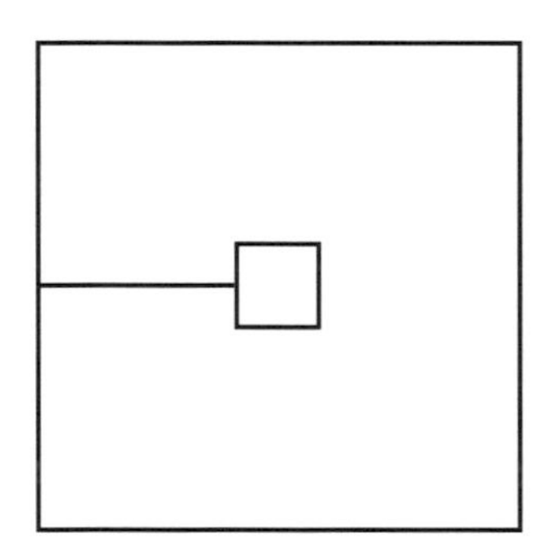

Fig. 18.4. Installer's solution

Problem 2

Students' Solutions. A student solution for this seam-placement problem is shown in figure 18.5. The students again used a paper floor plan and paper carpet pieces. The students found the length of the room to be eighteen feet and the width to be fourteen feet, and they chose to have the nap run in the direction of the width. With this arrangement, the students found that two pieces of carpet, each 12' by 14' 3" (considering trimming), are needed. After one of the pieces was placed, there was still a 6' 3" by 14' 3" section left to fill, so a second piece 12' by 14' 3" needed to be used. The students found that the total amount of carpet needed for this arrangement was 38 square yards (a 12' by 28' 6" piece of carpet); the cost was $910.10.

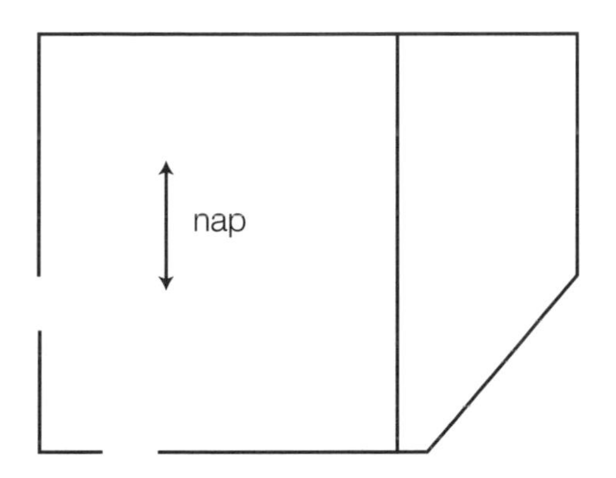

Fig. 18.5. Students' solution

Estimator's Solution. The estimator who encountered this problem figured the carpet installation and seam placement as shown in figure 18.6. The estimator tried the same arrangement as the students and also tried having the nap run in the direction of the length. In the latter case the estimator found that one piece of carpet 12' by 18' 3" and one piece of carpet 12' by 3' 8" were needed. The 12' by 18' 3" pieces of carpet would be laid in one piece. The 12' by 3' 8" piece would be cut into five pieces and laid at the end of the room away from the doors and out of the way of normal traffic patterns. The estimator determined the length of the shorter (3' 8") piece by dividing 12' (the width of the carpet) by the width of the space remaining after the large piece of carpet has been laid (2' 3"). The result of this calculation indicated that at most five pieces of carpet, each 2' 3" wide, can be cut from a piece 12' wide. Thus, there would be five carpet pieces used to fill the remaining space. Dividing the length of the room (18' 3") by the number of fill pieces (5) indicated that each fill piece would be 3' 8" in length (rounded up to the next inch). The nap of all the pieces of carpet would be running in the direction of the maximum length of the room. The total amount of carpet needed for this arrangement was 29.22 square yards (a 12' by 21' 11" piece of carpet); the cost of the carpet was $699.82.

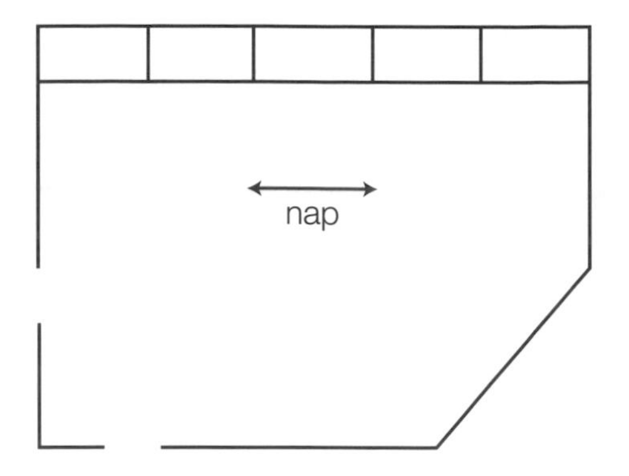

Fig. 18.6. Estimator's solution

PEDAGOGICAL CONSIDERATIONS

Having students work on constraint-filled problems such as these enables them to experience real-life problems and encourages them to bring their own ideas and knowledge to bear on the problems. Problems like these can be used as project problems during a measurement unit. For example, by combining these problems with additional problems set in contexts such as wallpapering and painting, students can investigate measurement concepts and processes through an interior-design project.

It is important to have follow-up discussions after problem-solving activities. Having students share their solutions and methods is a good way for students to see a variety of solutions to the problem. The discussion can focus on examining the solutions and deciding on a "best" solution(s). Of course, there may be more than one best solution, depending on the priority of different factors. In problem 1, the installer's solution is probably the best solution because it has the least amount of seam. In problem 2, to decide on a best solution one needs to weigh cost efficiency as well as seam placement. Considering both factors in the two solutions, the installer's is probably the better solution because it uses the lesser amount of carpet, and the seams (although more) are more out of the way of traffic patterns. As students gain experience with solving constraint-filled problems, they become better at finding optimal solutions.

SUMMARY

The teaching and learning of school mathematics needs to be connected to mathematics practice in everyday situations. At present, students are often not encouraged to make, and may even be discouraged from making, connections between everyday and school mathematics: "The process of schooling seems to encourage the idea that … there is not supposed to be much continuity between what one knows outside school and what one learns in school" (Resnick 1987, p. 15). Involving students in solving constraint-filled problems, such as the ones discussed here, engages them in doing mathematics in a meaningful context and encourages the understanding of the mathematical concepts and processes embedded in the problem and problem-solving process.

REFERENCES

Lester, Frank K., Jr. "Mathematical Problem Solving in and out of School." *Arithmetic Teacher* 37 (November 1989): 33–35.

Masingila, Joanna O. "Mathematics Practice and Apprenticeship in Carpet Laying: Suggestions for Mathematics Education." Doctoral diss., Indiana University, 1992.

National Council of Teachers of Mathematics. *Curriculum and Evaluation Standards for School Mathematics.* Reston, Va.: The Council, 1989.

———. *Professional Standards for Teaching Mathematics.* Reston, Va.: The Council, 1991.

Resnick, Lauren B. "Learning in School and Out." *Educational Researcher* 16 (December 1987): 13–20.

19

Mathematics and Quilting

Kathryn T. Ernie

MATHEMATICS and art design offer a rich environment for students to explore. Studying the works of M. C. Escher, who used transformations to create new tessellations and tiling problems, is a wonderful way to relate geometry and art. Because of a personal interest in history, art design, and mathematics, I have found quilting to be a fruitful area of study. The patchwork quilt, which might be thought of as one of the first recycling efforts of our early pioneers, presents a natural focus for studying geometric shapes and transformations. The history behind quilting, traditional quilt patterns, and the mathematics inherent in the quilt designs serve as interesting investigations. The patterns used in quilts were designed and given names that were meaningful to the period quilter. By their names, can you identify and visualize what these patterns would refer to: Anvil, Buggy Wheel, Churn Dash, Does and Darts, Eight Hands Round, and the Yankee Puzzle? (See fig. 19.1.) It is exciting to discover the history, literature, art, and mathematics of the American quilt. The mathematics of quilting demonstrates the strong connections between mathematics and art and between geometric concepts and algebraic concepts in an artistic setting.

QUILTS IN THE MATHEMATICS CLASSROOM

One way to incorporate the art of quilts in the classroom is to reenact a quilting bee as part of a school project. The historical study and preparation for the event should be an integral part of the process. It affords an excellent opportunity to discuss life on the Great Plains during the 1800s and why such occasions as quilting bees or barn raisings were important to American pioneers. Children's literature focusing on the quilt as a central theme is readily available. What a rewarding opportunity it is to share such award-winning multicultural books as *The Quilt* by Ann Jonas (1984), *Sweet Clara and the Freedom Quilt* by Deborah Hopkinson (1993), and *The Keeping Quilt* by Patricia Polacco (1988), among other stories. Lisa Campbell Ernst's (1983) book, *Sam Johnson and the Blue Ribbon Quilt,* will also encourage cooperation and participation by all children.

Fig. 19.1. Clockwise from upper left: Anvil, Buggy Wheel, Churn Dash, Does and Darts, Eight Hands Round, and the Yankee Puzzle

The mathematics explored by the class should include the design, construction, and measurement of the geometric shapes each child selects for his or her own patchwork. The actual project created by each student may be sewn or glued using fabric, or it could be created as an art-design poster using other materials: paper, paint, markers, stickers, or wood. A design may be inspired by traditional quilt patterns, or it may be a creative expression of the individual student.

For inspiration, let's consider a well-known Amish quilt pattern, often referred to as "Sunshine and Shadow" or "Trip around the World" (fig. 19.2). The pattern consists of rows of brightly colored squares, with each square a solid color. The squares would have been cut out of scraps of cloth saved by the quilter from such other sewing projects as shirts and aprons. Notice the two center lines (one vertical and the other horizontal) cutting the quilt into four regions. This is an example of folding symmetry.

One-fourth of the quilt, namely, the upper left corner (see fig. 19.3a), is the basic block used to create the quilt. This quarter of the quilt is reflected into the other three quadrants to create the rest of the design. The squares in a given row of the basic block are each a different color. To analyze the quilt, let's assign a different number to each unique color as it appears in this block. Starting in the top row, these could be numbered 0, 1, 2, 3, and so on, since each square is a different color. Now, using this color-to-number assignment, let's label the second row. Notice that the pattern is now: 1, 2, 3, 4, ... If we continue this process, the basic Amish block will be labeled as shown in figure 19.3b.

Fig. 19.2. Sunshine and Shadow quilt

Fig. 19.3a

0	1	2	3	4	5	6	7	8	9
1	2	3	4	5	6	7	8	9	0
2	3	4	5	6	7	8	9	0	1
3	4	5	6	7	8	9	0	1	2
4	5	6	7	8	9	0	1	2	3
5	6	7	8	9	0	1	2	3	4
6	7	8	9	0	1	2	3	4	5
7	8	9	0	1	2	3	4	5	6
8	9	0	1	2	3	4	5	6	7
9	0	1	2	3	4	5	6	7	8

Fig. 19.3b

What are some of the patterns we observe in this table? We see the diagonals that were apparent in the bands of color in the quilt. In addition, notice that a shift takes place between the rows in the table; it reminds us of an addition table, but with a difference. In this addition table we never use a number larger than nine. If the number is larger than nine, we subtract ten from it and use the difference in the table. Another way to describe this is to consider only the ones digits of the sums. This is an application of mod-10 addition, or "clock arithmetic." (Consider a clock labeled only with the hours 0 through 9. If the time is currently 5 o'clock on this mod-10 clock, what time will it be six hours from now? What time will it be 600 hours from now?)

Let's embed the basic quilt block into an addition table by adding the addends 0 through 9 to the left and top of the values, as shown in figure 19.4.

+	0	1	2	3	4	5	6	7	8	9
0	0	1	2	3	4	5	6	7	8	9
1	1	2	3	4	5	6	7	8	9	0
2	2	3	4	5	6	7	8	9	0	1
3	3	4	5	6	7	8	9	0	1	2
4	4	5	6	7	8	9	0	1	2	3
5	5	6	7	8	9	0	1	2	3	4
6	6	7	8	9	0	1	2	3	4	5
7	7	8	9	0	1	2	3	4	5	6
8	8	9	0	1	2	3	4	5	6	7
9	9	0	1	2	3	4	5	6	7	8

Fig. 19.4

Consider the properties of this finite system as defined by the elements {0, 1, 2, 3, 4, 5, 6, 7, 8, 9} and mod-10 addition. This is an example of a closed system; adding any two of the elements using mod-10 addition results in one of the elements of the set, as can readily be seen from the table. What other properties are apparent from the study of this finite system? Is mod-10 addition commutative over this finite set? Is it associative? Does this system have an additive identity? Compare this finite system with the addition of whole numbers. If the finite system is changed from mod-10 to a different mod-n, $n > 1$, will the same properties hold?

What will happen if a different operation is used, such as subtraction or multiplication? How will this change affect the overall artistic design of the basic block? Notice that subtraction will be quite similar, the difference being that the diagonals of the same color (same number) will run from the upper left to the lower right, the opposite of the addition table. The complete mod-5 subtraction table over the set {0, 1, 2, 3, 4} is shown in figure 19.5a.

Again a clock analogy is useful in determining the results of mod subtraction. To accomplish subtraction, move the number of hours subtracted in the counterclockwise direction to obtain the result. For mod-5, create a clock with the five values 0, 1, 2, 3, and 4, as the hours on the face. To determine (2 – 4) in mod-5, start at 2 and move counterclockwise four "hours." The resulting difference is 3.

Compare the finite system defined by mod-5 subtraction over the set {0, 1, 2, 3, 4} with the finite system defined by mod-10 addition. Do the same properties hold for both systems? Now, using a one-to-one correspondence assigning a unique color or pattern to each of the five elements in the finite set, the student can create a block based on mod-5 subtraction. An example is shown in figure 19.5b. Can you identify the transformations used to create each of the four quilts in figure 19.6 using mod-5 subtraction in the basic block?

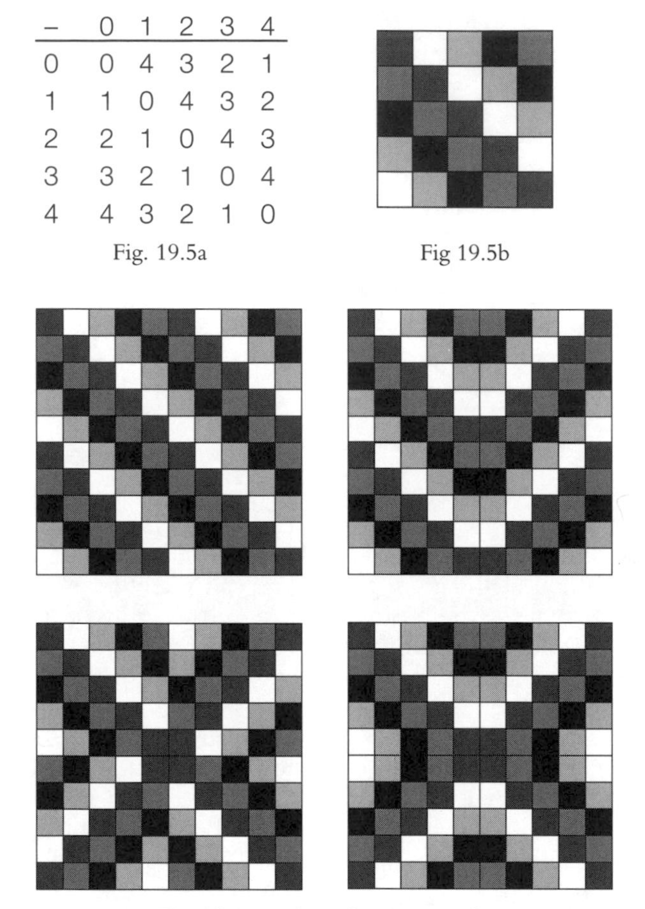

–	0	1	2	3	4
0	0	4	3	2	1
1	1	0	4	3	2
2	2	1	0	4	3
3	3	2	1	0	4
4	4	3	2	1	0

Fig. 19.5a

Fig 19.5b

Fig. 19.6. Mod-5 subtraction quilts

TECHNOLOGY

The author was able to create the quilt patterns discussed in this article using the Claris CAD (1992) drawing program on a Macintosh Centris 660AV. Using drawing software, students can experiment as they create the basic quilt blocks. They can create various geometric shapes for each of the elements and even experiment with the effect of color on the overall design. Once the basic quilt block is completed, students can copy the block and then transform the basic block as they create the final quilt project. Most drawing systems have built-in commands for translating, rotating, and reflecting objects as a whole. This experimentation allows students to study the effects of different mod systems, operations, color, basic element designs, and transformations much more quickly than if they had to do all this initial design work by hand. Once the final quilt is developed on the computer, the students can share the development process with their peers by demonstrating the different stages of their quilt projects, saved on disk, using either a monitor or printouts. The final design may then be implemented in another medium such as on poster board or with cloth.

CONCLUSION

The examples in this article illustrate how students can use modular arithmetic and transformational geometry to create quilt designs. Students may also want to study and use other finite structures based on other operations to create their own designs. The cell designs used in the cell-to-number assignment can be selected to tell something about the student and his or her history. Each cell in the basic block can be constructed from geometric shapes other than a square. Common selections include triangles, arches, and rectangles.

Students also can analyze the mathematics involved in other quilt patterns. However, it should be noted that not all quilts make use of transformations of a basic design. (For example, consider the randomness of a crazy quilt or a quilt that uses one large design or appliqué.) Exciting problem solving takes place as students analyze quilts for the patterns, symmetry, mathematical elements, and structure.

Although we have focused on the American patchwork quilt, a natural extension is to study the mathematics of other forms of quilting and to discover the influence of other cultures on this art form. Clearly, the subject of quilting furnishes many opportunities for connections within the school curriculum.

BIBLIOGRAPHY

Bezuszka, Stanley, et al. *Designs from Mathematical Patterns.* Palo Alto, Calif.: Dale Seymour Publications, 1990.

Burton, Grace, et al. *Fourth-Grade Book. Curriculum and Evaluation Standards for School Mathematics* Addenda Series, Grades K–6, edited by Miriam A. Leiva, pp. 9–11. Reston, Va.: National Council of Teachers of Mathematics, 1992.

Carey, Deborah A. "The Patchwork Quilt: A Context for Problem Solving." *Arithmetic Teacher* 40 (December 1992): 199–203.

Claris CAD. Version 2.0.3. Santa Clara, Calif.: Claris Corporation.

De Temple, Duane. "Reflection Borders for Patchwork Quilts." *Mathematics Teacher* 79 (February 1986): 138–43.

Ernst, Lisa Campbell. *Sam Johnson and the Blue Ribbon Quilt.* New York: Mulberry Books, 1983.

Flourney, Valerie. *The Patchwork Quilt.* New York: Penguin Books, 1985.

Forseth, Sonia, and Andria Price Troutman. "Using Mathematical Structures to Generate Artistic Designs." *Mathematics Teacher* 67 (May 1974): 393–98.

Haders, Phyllis. *Sunshine and Shadow: The Amish and Their Quilts.* New York: Universe Books, 1976.

Hopkinson, Deborah. *Sweet Clara and the Freedom Quilt.* New York: Alfred A. Knopf, 1993.

Jenkins, Susan, and Linda Seward. *The American Quilt Story.* Emmaus, Pa.: Rodale Press, 1991.

Jonas, Ann. *The Quilt.* New York: Greenwillow Books, 1984.

National Research Council, Mathematical Sciences Education Board. *Measuring Up: Prototypes for Mathematics Assessment,* pp. 85–94. Washington, D.C.: National Academy Press, 1993.

Polacco, Patricia. *The Keeping Quilt.* New York: Simon & Schuster, 1988.

Zaslavsky, Claudia. "Symmetry in American Folk Art." *Arithmetic Teacher* 38 (September 1990): 6–12.

20

Randomness: A Connection to Reality

Donald J. Dessart

RANDOMNESS surrounds us in everyday living. The results of a presidential poll, the toss of a coin, the winning numbers in a lottery, the spinning of a spinner are a few examples of common random phenomena. Some important questions are these:

- What is randomness?
- What do students think about randomness?
- What are some classroom activities related to randomness?

WHAT IS RANDOMNESS?

Randomness is an elusive concept. In textbooks, it is often left undefined, like the concept of *point* in geometry (Kolata 1986, p. 1068). Moore (1990, p. 98) gave this description: "Phenomena having uncertain individual outcomes but a regular pattern of outcomes in many repetitions are called *random*." A random phenomenon, therefore, has two distinct characteristics: (1) one cannot predict the outcome on a single occurrence of the phenomenon, but (2) there is a pattern in many repetitions of the phenomenon. For example, one cannot predict heads or tails on a single toss of a coin, but if the coin is tossed a large number of times, about one-half of the tosses will be heads and about one-half will be tails.

Randomness entered into a celebrated case of alleged scientific fraud, the "Baltimore Case" (Hamilton 1991, p. 1168). In a subcloning experiment, the quantity of antibody present in a solution was determined using a radioactive label, such as iodine-125, and measuring the number of gamma decays with a radiation counter. The investigator copied some of the radiation-count data by hand. If the recorded data had been generated by a radiation counter, then digits in the tens column should have been random. In 315 counts, there appeared to be an overabundance of

1s and 3s in the tens column, leading to the speculation that the data may have been fabricated. The actual data set is given here:

Integer	0	1	2	3	4	5	6	7	8	9
Frequency of occurrence	14	71	7	65	23	19	12	45	53	6

Since the total frequency is 315, one might expect that each of the ten digits should appear about thirty times, but *1* occurred seventy-one times and *3* occurred sixty-five times! The data were subjected to a chi-square analysis, which revealed that such a distribution had one chance in 10^{32} of occurring randomly. How large is 10^{32}? Since the radius of the entire universe was estimated by Einstein to be 10^{26} feet, one realizes the immensity of 10^{32} and the high improbability, but not impossibility, that the set was generated by a radiation counter. When confronted with this nonrandom possibility, the investigator did not offer a ready explanation.

WHAT DO STUDENTS THINK ABOUT RANDOMNESS?

Green (1987) studied children's understanding of randomness and reported on a survey of children aged seven to fourteen in England concerning raindrop patterns. That survey was repeated by Dessart (1991) with middle school children and first-year college students in Tennessee. The middle school children were aged ten to thirteen, and the college students were in a freshman calculus class. The students were given the following problem:

> The flat roof of a small building has sixteen square sections. It begins to rain. After sixteen raindrops have fallen, which of these three patterns would you expect to see?

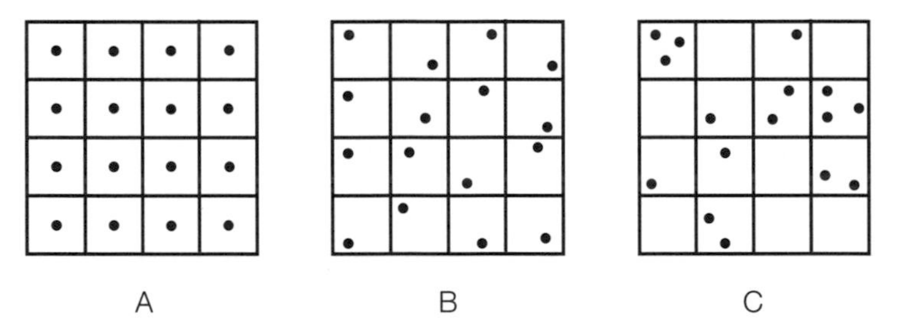

Pattern A is called the "ordered case." In this case, a single raindrop falls exactly in the center of each of the sixteen squares, which makes the pattern of raindrops highly predictable. In Pattern B, the "semiordered case," a raindrop falls in each of the sixteen squares but not at the center of the square. The position of each drop is not as easily predicted but nevertheless is predictable. Pattern C is called the "random case," since the pattern of drops does not appear to be predictable.

In both England and Tennessee, the teachers were asked to subjectively classify their students by reasoning-ability levels as *A, B, C, D,* or *E* in England and as *high, average,* or *low* in Tennessee. The groups were not entirely comparable, and rigorous statistical controls cannot be employed in these comparisons; nevertheless, the information revealed in table 20.1 is enlightening.

TABLE 20.1
Percent of Responses to the Raindrop Pattern

Students	Pattern A	Pattern B	Pattern C	None or Other
High ability U.S.A.	3	49	40	8
Average ability U.S.A.	10	44	38	8
Low ability U.S.A.	16	42	38	4
A ability England	8	40	42	11
C ability England	15	36	37	12
D ability England	17	36	33	14
College U.S.A.	0	52	31	17

As one can see from table 20.1, the ordered case, Pattern A, in which a single raindrop fell in the center of each of the squares, was selected by 16 and 17 percent of the lower-ability children in each country and by only 3 and 8 percent of the higher-ability children. None of the U.S. college students chose Pattern A. The random case, Pattern C, was selected by 33 and 38 percent of the lower-ability children, 40 and 42 percent of the higher-ability children, and 31 percent of the college students. The semi-ordered case, Pattern B, varied from 36 percent for *C* and *D* children in England to 52 percent of the college students in Tennessee.

These statistics revealed that students were more likely to perceive the patterns of raindrops as ordered rather than random. Consequently, teachers should help students see that randomness implies that a particular instance of a phenomenon is nonpredictable but that there is a pattern in many repetitions of the phenomenon. Most surprising in this study was that there were no marked differences between children and college students. In fact, the remarkable similarity in the choice of Pattern C by low-ability students in England and college students in the United States reinforces the notion that randomness is both an elusive and bewildering concept.

CLASSROOM ACTIVITIES

The generation of random numbers, the testing for randomness, and the use of random numbers in simulation may serve as classroom activities.

Generation of Random Numbers

A generator that students can use directly with a basic, hand-held calculator or that can be easily programmed on a computer or programmable calculator is described below:

1. Select any seven-digit number preceded by a decimal point and with the last digit odd and other than 5, for example, 0.276 543 3.
2. Multiply the decimal obtained in step 1 by 147, that is, $0.276\,543\,3 \times 147 = 40.651\,865\,1$.
3. From that product subtract the whole-number portion of the decimal, that is, $40.651\,865\,1 - 40 = 0.651\,865\,1$.
4. Continue this process by repeating steps 2 and 3 (multiplying by 147 and subtracting the whole-number portion of the product) to obtain a sequence of random numbers.

One-digit random numbers are represented by the first digits of the numbers generated by the "147 algorithm" shown above; two-digit random numbers by the first two digits; three-digit random numbers by the first three digits; and so on (Rade 1981).

In a classroom, teams of students might compare the generation of random digits using a 0–9 spinner to those generated by the 147 algorithm. The students should take turns recording, calculating, or spinning. Students see that over many trials the frequency of each of the digits becomes more nearly equal as the number of trials grows.

A Simulation Problem

Random numbers can be used to solve difficult problems by simulation. Simulation depends on the fact that if an experiment is repeated a large number of times, the relative frequency of an observed phenomenon settles about a fixed value associated with that phenomenon.

> ***The Bottle-Cap Problem.*** Lison Dairy decides to initiate a contest to encourage customers to purchase more milk. One of the letters *L, I, S, O,* or *N* is printed on the inside of each bottle cap. The same number of each letter is printed, and the caps are thoroughly mixed before the bottles are capped. Any customer obtaining enough caps to spell *LISON* wins a prize. How many caps would one expect to have to obtain to win a prize?

This problem can be simulated with random numbers by using the correspondence $L \leftrightarrow \{0, 1\}$; $I \leftrightarrow \{2, 3\}$; $S \leftrightarrow \{4, 5\}$; $O \leftrightarrow \{6, 7\}$; $N \leftrightarrow \{8, 9\}$. Each student simulates the number of purchases needed before collecting all five letters. The class data can also be combined and a mean, or expected number of purchases, calculated. The expected number should be approximately 11.5 (Lappan and Winter 1980). (For a detailed discussion of a similar problem, see the article by Schielack in this yearbook.)

CONCLUSION

Randomness is a connection to reality that touches life in many ways. Lotteries, polls, and children's games often depend on a random device. Simulation represents a powerful means of solving difficult problems.

REFERENCES

Dessart, Donald J. "Randomness: Thoughts of Students." Unpublished raw data, 1991.

Green, David R. "Children's Understanding of Randomness: Report of a Survey of 1600 Children Aged 7–11 Years." In *Proceedings: The Second International Conference on Teaching Statistics,* edited by Roger Davidson and Jim Swift, pp. 287–91. Victoria, British Columbia: University of Victoria, 1987.

Hamilton, David P. "Verdict in Sight in the 'Baltimore Case.'" *Science,* 8 March 1991, pp. 1168–72.

Kolata, Gina. "What Does It Mean to Be Random?" *Science,* 7 March 1986, pp. 1068–70.

Lappan, Glenda, and M. J. Winter. "Probability Simulation in Middle School." *Mathematics Teacher* 73 (September 1980): pp. 446–49.

Moore, David S. "Uncertainty." In *On the Shoulders of Giants: New Approaches to Numeracy,* edited by Lynn Arthur Steen. Washington, D.C.: National Academy Press, 1990.

Rade, Lennart. "Random Digits and the Programmable Calculator." In *Teaching Statistics and Probability,* 1981 Yearbook of the National Council of Teachers of Mathematics, edited by Albert P. Shulte, pp. 118–25. Reston, Va.: The Council, 1981.

PART 5

Connections across the High School Curriculum

21

Connecting Geometry with the Rest of Mathematics

Albert A. Cuoco
E. Paul Goldenberg
June Mark

THE beauty of mathematics lies largely in the interrelatedness of its ideas. For the mathematician, making these interconnections not only displays that beauty but also generates new research techniques. If students can make these connections, will they also see beauty in mathematics? We think so, and in this article, we illustrate one approach made practical by a new breed of geometry software—an approach that blends experimental investigation of problems with theoretical analysis.

The Geometer's Sketchpad (1990), Cabri, Cabri Geometry II (1992), Tangible Math: Geometry Inventor, and the Geometric superSupposer (1992) allow students to build dynamically deformable geometric constructions and to study how some feature of a construction behaves under the deformation (see Klotz's [1991] early description of Sketchpad's features for a characterization of these dynamic geometry software tools.). These tools are typically used in the study of classical geometry, but a change of perspective offers opportunities to connect geometric thinking with the rest of mathematics. Dynamic constructions are functions in which one geometric feature is the independent variable and another feature is the function's value. These functions, defined on and taking values in R or R^2 (the real numbers or real plane), are specified by geometric

The activities described here are from a module on optimization that is part of Education Development Center's Connected Geometry project, supported by the National Science Foundation, grant number MDR-9252952. Additional support was received from EDC development funds, NSF grants numbers MDR-8954647 and MDR-9054677, and Apple Computer, Inc., External Research Division. The views represented here are not necessarily shared by any of the funders. We thank Wayne Harvey and Glenn Kleiman of EDC and James Sandefur of NSF for their contributions to the ideas presented here.

constructions rather than by algebraic formulas. Analyzing the behavior of these functions leads students to connect their knowledge of geometric facts with major themes like continuity, optimization, and limit. Software tools also supply environments in which students can experience mathematical research; by using the software to investigate functions defined on geometric objects, students can feel the interplay between experiment and proof that is central to mathematical research.

We can illustrate these ideas with an investigation that has been used successfully with students of widely differing backgrounds. The investigation requires some geometric facts; at the teacher's discretion, the activities may be used either to apply these facts or to develop them.

A GEOMETRY PROBLEM FOR STUDENTS TO INVESTIGATE

> Three cities get together and decide to build an airport. Where should the airport go?

With discussion, students often raise varied approaches, each worth following up. The one pursued here is the *environmental solution:* The placement should minimize total travel to and from the airport. For cities of roughly equal population, this minimizes the sum of the distances from the airport to the cities.

This solution leads to an investigation of a problem attributed to Fermat:

> Given a triangle, how can one locate a point P to minimize the sum of the distances from P to the vertices of the triangle?

Investigating this question establishes connections among a great many topics from high school mathematics, ranging from first-year algebra to precalculus. Our approach also fosters explicit discussions of important mathematical methods such as reasoning by continuity, the dialectic between proof and experiment, and the heuristics of calculating area in more than one way and of relating problems that have a similar "feel." Here is one path through an investigation of the airport problem.

Stage 1: An Experimental Approach

Students begin with a physical manipulative like the one shown in figure 21.1. Nails or pins driven into a board represent the cities. With fishing line laced as shown, students can simply pull and see where the ring is drawn.

Experiments with mechanical devices serve three purposes. First, by embodying notions like distance and minimizing distance in the concrete act of "pulling out the slack," students achieve an intuitive sense for problems of this kind. Second, such mechanical devices build connections between geometry and the world of levers, hinges, bouncing balls, and so on. These connections help students bring their common sense to geometry and equally help students bring geometric insights back to applications in

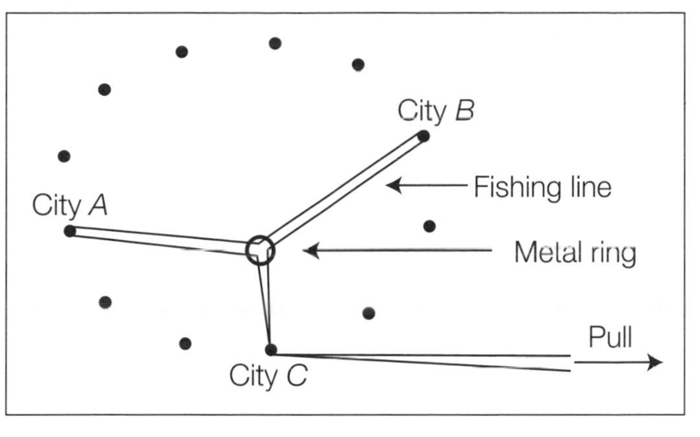

Fig. 21.1

the physical world. Most important, this experiment opens a discussion about whether or not there is a *single* point at which the minimum distance occurs. The manipulative makes this seem likely, so students are now on the lookout for a point rather than a nebulous region of solutions.

By relocating the vertices, students can try several configurations of cities and develop a collection of conjectures that can be compiled and refined through class discussion. Some of the conjectures may be incorrect, incomplete, or vague:

1. "The best spot is at the middle."
2. "The airport will be the same distance from all cities."
3. "The airport should go on one of the cities."

Others will be on the right track, with varying degrees of precision:

4. "If two of the cities are very close to each other and far from the third city, the airport will be closer to the first two."
5. "For most arrangements, the airport winds up sort of in the middle. We measured the angles around the airport, and they are all about the same. We conjecture they should be exactly the same: 120°."
6. "If every angle of the triangle measures less than 120°, then the sum of the distances is minimized at a point around which the airport roads make angles of 120°. But if one angle of the triangle is larger than 120°, the best spot for the airport is at the vertex of the largest angle."

The higher-numbered conjectures above are unlikely in a high school class; students seldom phrase their ideas in formal ways and often need help in communicating their conjectures more precisely.

To test their conjectures, students can build modifiable constructions on a computer. Suppose, for example, that students are testing out conjecture

5 above. One way to locate their hypothetical "best spot" (the place that makes 120° angles with the triangle's vertices), is to use a setup like the one shown in figure 2. Here, the fixed whiskers $\overline{DE}$ and $\overline{DF}$ make 120° angles with $\overline{AD}$. Students can move *D* until $\overline{DC}$ and $\overline{DB}$ lie along these conjecturally best directions. They can then record the sum of the distances and compare it with the sum at other positions for *D*. By dragging vertices *A*, *B*, or *C* around, they can reproduce the experiment on many different triangles.

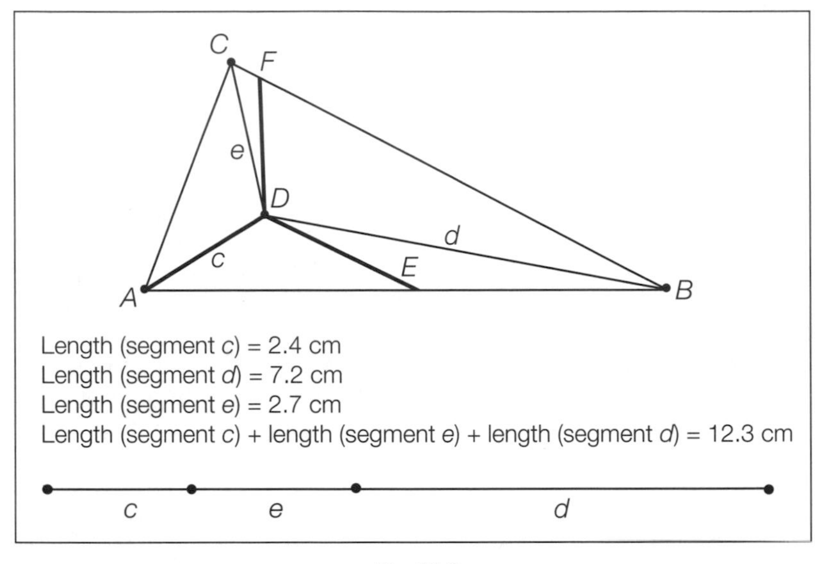

Fig. 21.2

Time spent on this experiment gives students a chance to gather data to support their conjectures and serves the important purpose of allowing students to experience continuity properties of functions defined on geometric constructions. They are experimenting with a function, call it *A*, from points on the plane to the real numbers. As the input to that function (point *D*) is moved, the student can see the change in the function's value, *A*(*D*), numerically and as a dynamically changing line. Teachers have been trying for years to use pictures on the chalkboard or animated films to conjure up this image of a smoothly changing system. This experiment puts the variable to such a system in students' hands.

Students should not get the impression that mathematical results are established solely on the basis of data-driven discovery. The experiment should also be viewed as an opportunity to discuss the need for theoretical justification in the face of experimental error and to emphasize the fact that the search for a proof of a conjecture often leads to new and surprising results and connections.

This stage might well close, then, with a list of conjectures posted on the board, a discussion of why these conjectures need revisiting, and some computer experiments that can be modified to gain insights into other problems. The investigation can be reopened some time later, perhaps while studying equilateral triangles.

Stage 2: A Related Problem

High school students love to hear stories about other high school students, especially if the stories involve outsmarting adults. (One of the authors still uses the story of how a former student, now out of high school for two decades, found a four-line proof of a fact in matrix algebra that had previously taken several chalkboards to establish. His story now enchants students young enough to be his children.) Several years ago, a student, Richard, was faced with a problem on a standardized test in which he was given an equilateral triangle of side 10 and a point D in the triangle's interior; he was asked to find the sum of the distances from D to the sides of the triangle (fig. 21.3).

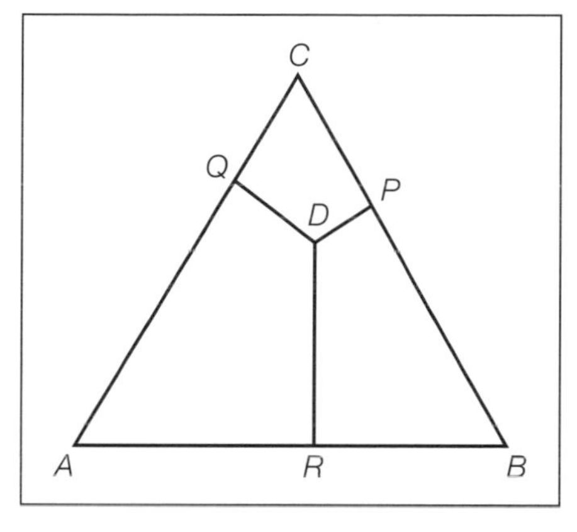

Fig. 21.3

Not knowing any special facts, Richard reasoned that since the problem didn't say anything special about D, he could move it anywhere he liked. So, he moved D very close to vertex C, where two of the distances were almost 0, and the other was almost an altitude of the triangle. He reasoned that the answer to the question must therefore be the length of the triangle's altitude, which he could easily find.

Richard assumed that the function, call it R, defined on the triangle and its interior as $R(D) = DP + DQ + DR$, is continuous and constant. His assumption that the function is constant came from the environment of a standardized test, but his assumption that it is continuous came from his ability to do the thought experiment of "moving the point around" inside the triangle, seeing that small perturbations to D produce small changes in the sum of the distances. It is not surprising that most students do not realize that the function is constant, but many of them also have a hard time visualizing the dynamics that Richard carried out in his head. Using appropriate software, students can design their own experiments and develop the kind of reasoning by continuity that Richard used.

After setting up the experiment (see fig. 21.4), students can move *D* within the triangle, seeing the effect that *D*'s position has on the sum of the distances *and* on the individual contributions of *DQ*, *DR*, and *DP* to the sum. The continuity of this function becomes apparent as soon as one starts dragging *D* around, but the surprise is that although the individual contributions of *DQ*, *DR*, and *DP* change as *D* moves, their sum seems to stay constant. A conjecture emerges, and this time it is open to analysis.

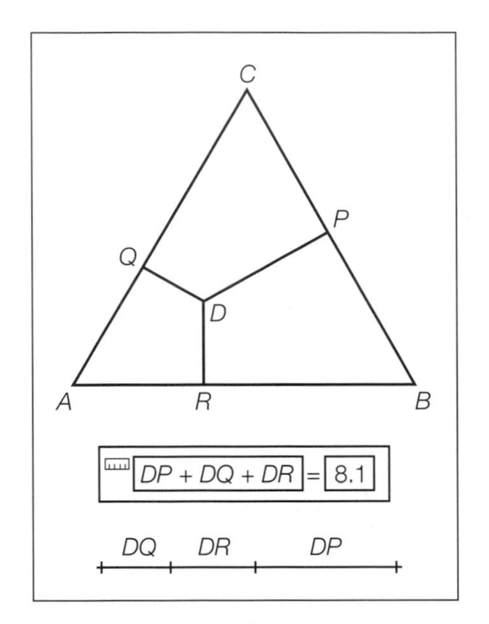

Fig. 21.4

There are at least two ways to approach the analysis:

- Using the area formula for triangles and elementary algebra
- Using a geometric thought experiment inspired by the April, 1992 cover of the *Mathematics Teacher*

Space limitations prohibit looking at the second approach (details are contained in Connected Geometry's module "Optimization"), but let's look at a connection between the geometry of Richard's function and first-year algebra.

A powerful heuristic approach in geometry is to calculate the area of a figure in more than one way and to compare the results. On the first trial, students need to see an example of this technique, and the equilateral-triangle theorem is a good context in which to do it. If the side length of the triangle is s and its altitude has length h, then the triangle has area $(1/2)sh$. Conversely, if we draw in "airport lines" to the vertices (fig. 21.5), we see that the triangle's area can be found as the sum of the areas of three smaller triangles, so that we can calculate:

$$\begin{aligned}\frac{1}{2}sh &= \frac{1}{2}s \cdot DP + \frac{1}{2}s \cdot DQ + \frac{1}{2}s \cdot DR \\ &= \frac{1}{2}s\left(DP + DQ + DR\right)\end{aligned}$$

Therefore, $DP + DQ + DR = h$, and the sum of the distances to the sides is equal to the length of an altitude of the triangle.

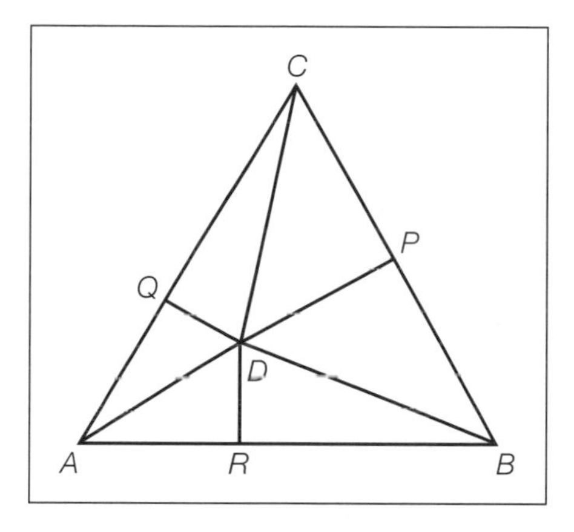

Fig. 21.5

This proof could simply be presented by the teacher. Alternatively, the teacher can write it out, ask a group of students to study it, and then have that group present the proof to the class. Still another approach is for the teacher to lead a discussion in which a general proof is treated as if it were restricted to a special case (e.g., a case in which, say, the length of the side of the triangle is 10), and then to ask students to construct the general proof. Important discussion questions, after the class has seen the proof, are the following:

- Why must the triangle be equilateral for the proof to work?
- If the triangle is not assumed to be equilateral, the proof fails. Does that mean that the fact is no longer true?

A Generalization

Suppose that the triangle is *not* equilateral (fig. 21.6). Students can modify any of their previous experiments and see that the sum of the distances from *D* to the sides of the triangle is *not* constant: As *D* is moved around the triangle's interior, the sum varies along with its component pieces.

The following are natural questions generated by this modification: Where is the sum of the distances smallest? Where is it largest?

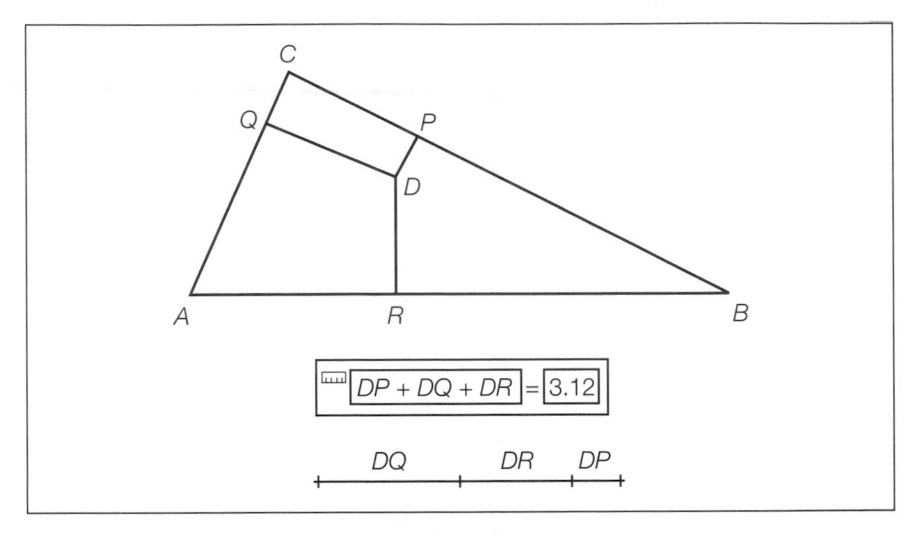

Fig. 21.6

Experimentation easily leads to the conjecture that the minimum distance occurs at the vertex opposite the longest side (the vertex with the greatest angle) and the maximum occurs at the vertex opposite the shortest side. Students can look at their proofs for the equilateral case and modify them to establish this new conjecture. For example, if the minimum occurs at *C*, then it must be the length of the shortest altitude of the triangle. Call that height *h*. Similarly, if the maximum occurs at *B*, it must be the length of the longest altitude of the triangle. Call *that* height *H*. (See fig. 21.7.)

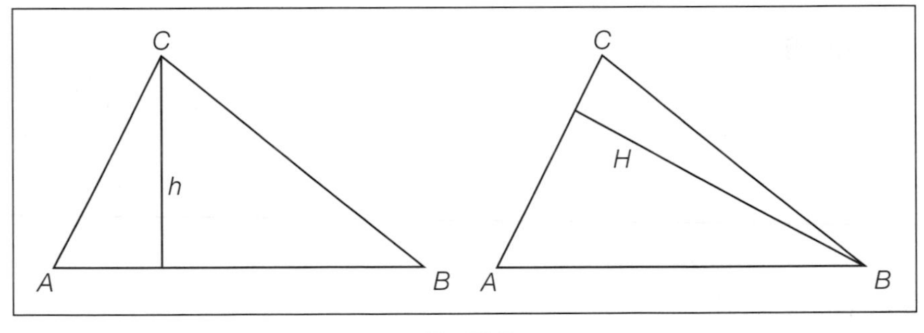

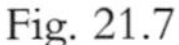

Fig. 21.7

Then, for any point *D* in this triangle's interior, we must show that the sum of the distances from *D* to the sides lies strictly between *h* and *H*.

As before, we can divide the triangle into three smaller triangles and add up the areas (fig. 21.8). This leads to a slight variation of our earlier calculation:

$$\frac{1}{2}AB \times h = \frac{1}{2}AC \times DQ + \frac{1}{2}CB \times DP + \frac{1}{2}AB \times DR$$

$$< \frac{1}{2}AB \times DQ + \frac{1}{2}AB \times DP + \frac{1}{2}AB \times DR$$

(because $\overline{AB}$ is the longest side)

$$= \frac{1}{2}AB(DQ + DP + DR)$$

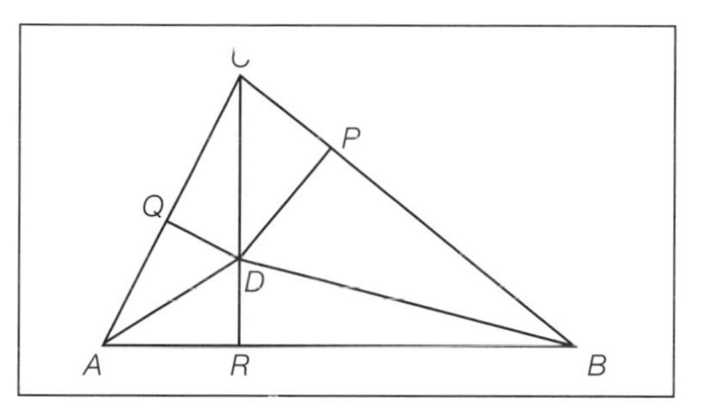

Fig. 21.8

This implies that $h < DQ + DP + DR$, as desired. Arguing similarly, we see that $H > DQ + DP + DR$. Students encounter a situation where choosing to use an inequality helps them to make progress—an important connection.

A Discussion Topic

We have seen that if D is a point in the triangle's interior, then $h < R(D) < H$ where h and H are respectively the lengths of the shortest and longest altitudes. What happens if h gets very close to H? First, the triangle must be close to equilateral (because all three of its altitudes would have roughly the same lengths); second, $R(D)$ would be sandwiched between two numbers that are nearly equal. Passing to the limit, we can use our more general result to recapture the earlier theorem about equilateral triangles: The sum of the distances from any interior point to the sides of an equilateral triangle equals the length of the triangle's altitude.

This activity furnishes an opportunity for a class discussion on mathematical method. So often in mathematics, a result is noticed in a special case, and an explanation is constructed that makes essential use of the particular features of the special case. As more general cases are considered, experimentation shows how the original explanation must be modified to accommodate the new generality. New conjectures arise. Then the actual methods used to establish the special result can be studied, and with a little perseverance, one can modify the arguments to establish the more general conjectures. These general theorems, in turn, imply the original result that stimulated the whole investigation, offering deeper insights into it and placing it in a broader context.

Stage 3: The Airport Problem Revisited

In figure 21.9, $\overline{DA}$, $\overline{DB}$, and $\overline{DC}$ are the three segments used to compute the airport function for $\triangle ABC$. The goal is to find a location for D that minimizes that function. The class has long had experiment-based conjectures, but as yet, no method of analysis.

The teacher reminds the students that although they know little about sums of distances to vertices, they know quite a bit about sums of distances to sides. The teacher might prompt: "Well, of course we could *make* these very same segments be distances to the sides of a triangle by drawing perpendiculars to the airport roads at cities A, B, and C, like this." (See fig. 21.10.)

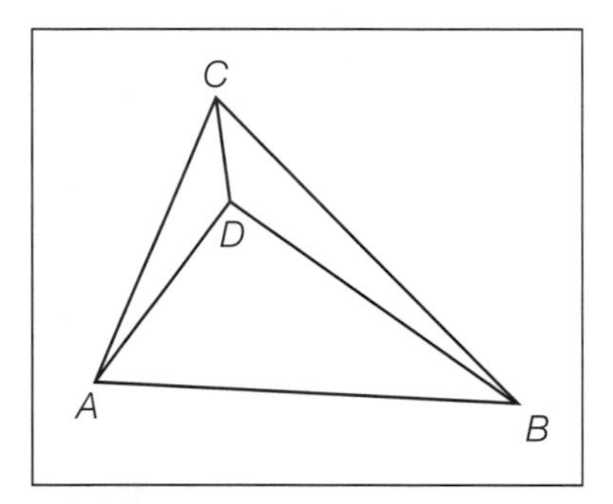

Fig. 21.9

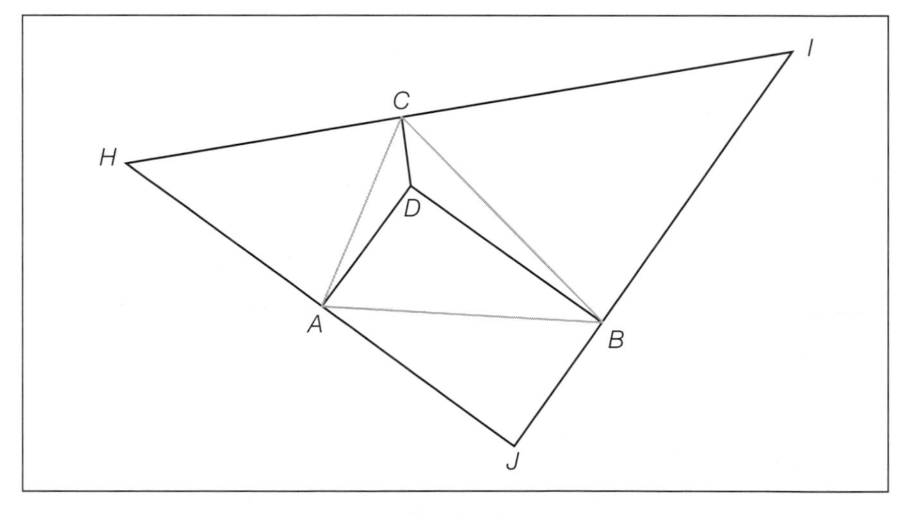

Fig. 21.10

The construction is easy to effect with the software. What happens to the outer triangle when the airport is moved to its conjectured best spot (where all central angles are 120°)? A little work and some results from classical geometry show that the big triangle must then become equilateral (fig. 21.11).

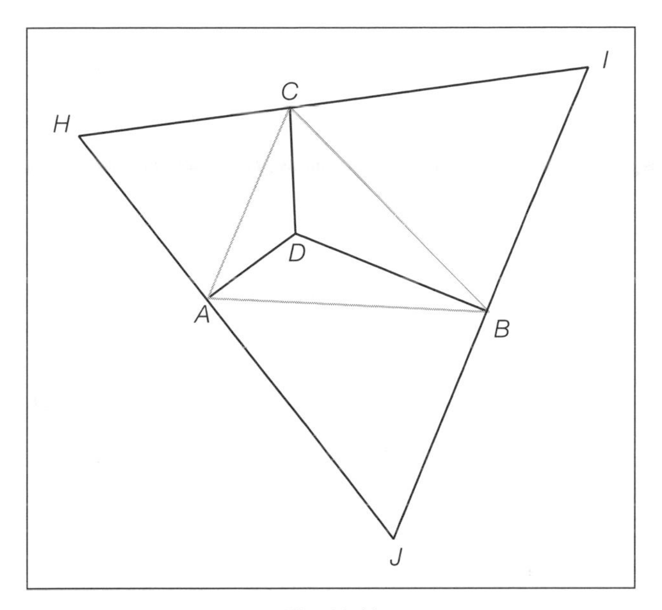

Fig. 21.11

Now we can apply what we've learned about equilateral triangles. Because the big triangle is equilateral, the sum of the distances from any interior point S to its sides, a number we're calling $R(S)$, is equal to the triangle's altitude, constant on the big triangle's interior: in particular, $R_{\triangle HIJ}(S) = R_{\triangle HIJ}(D) = A_{\triangle ABC}(D)$. So we pick some point S (other than D) and compare $R_{\triangle HIJ}(S)$ with the sum $SA + SB + SC$, S's distances to the vertices of $\triangle ABC$. The shortest distance from a point to a line is a perpendicular distance, and so $R_{\triangle HIJ}(S)$ (which is the same as $R_{\triangle HIJ}(D)$) is smaller than $SA + SB + SC$. That is, $DA + DB + DC < SA + SB + SC$. So D is the point that minimizes the airport function!

Extensions

There are many directions for further investigations:

1. The construction explained above fails if it is impossible to locate a point D inside $\triangle ABC$ in a way that makes $\triangle HIJ$ equilateral. As the largest angle of $\triangle ABC$ approaches 120°, the airport moves quite clearly toward that vertex. Why?

2. We have been looking at two functions: R measures the sum of the distances from a point to the sides of a triangle, and A, the airport function,

measures the sum of the distances to the vertices of a triangle. These functions, defined on the entire plane instead of merely on the interior of the triangle, are ideal objects for study. For one example, what is the locus of all the points for which *A* produces some constant value *k?* Using analytic geometry, students can choose a triangle on the Cartesian plane and, using the coordinates of its vertices, obtain an expression for *A* in either Cartesian or vector form. For example, if the vertices are (0, 0), (6, 0), and (2, 8), students will be faced with graphing equations of the form

$$\sqrt{x^2 + y^2} + \sqrt{(x-6)^2 + y^2} + \sqrt{(x-2)^2 + (y-8)^2} = k.$$

This looks like the unsimplified equation of an ellipse, except we sum three distances instead of two. Jack Janssen, a Connected Geometry staff member, realized that the pin-and-string construction for an ellipse can be generalized to obtain curves described by equations like these using the "airport board" (fig. 21.12), suggesting that the curves be considered a generalization of conics. Several computer systems graph equations of this type. Mathematica (1991), for example, will graph an entire family of such curves, producing a "contour plot" of the function *A* (fig. 21.13). By showing lines of constant value, contour plots furnish us with a kind of topographic map of a function,

3. Exploring the contour lines for *R* shows how there is no one "best" way to visualize a function. In figure 21.14, we see an equilateral triangle of side 10 on the Cartesian plane. Using the formula for the distance from a point to a line, we constructed an expression for *R* and used Mathematica to plot its contour lines. The curves in the graph result from inaccuracies in the graphing program, but even this picture is sufficient to help us conjecture that for an equilateral triangle, the contour lines are hexagons whose vertices lie on the extensions of the triangle's sides and

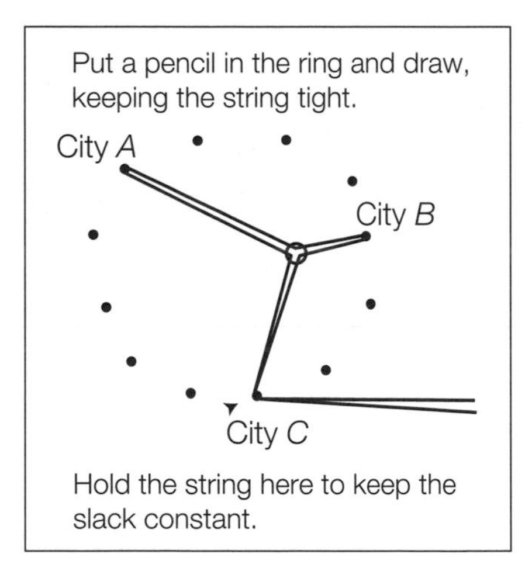

Fig. 21.12

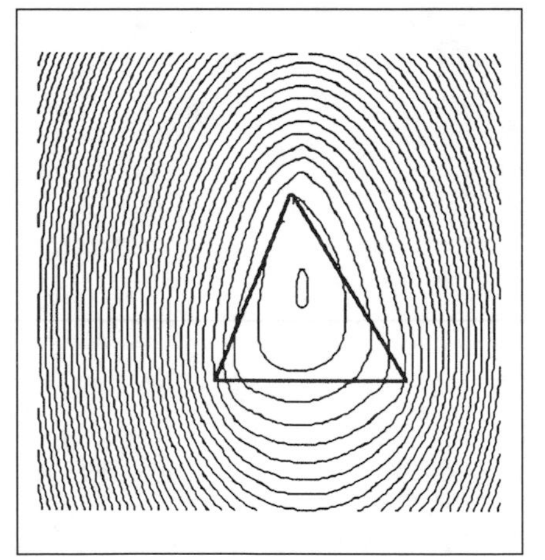

Fig. 21.13

whose sides are parallel to those of the triangle. Having come to this conjecture experimentally, we see that it is open to analysis through a combination of geometric reasoning and reasoning by continuity (fig. 21.15).

But what happens in the case of a nonequilateral triangle? We've seen that the minimum value is at the vertex opposite the longest side and the maximum is at the vertex opposite the shortest side. A contour plot (fig. 21.16) supports this observation but is otherwise quite mysterious.

Another kind of graph helps solve the mystery. Just as topographic and relief maps are two ways to model the same terrain, contour and surface plots are two ways to visualize a function from the plane to R. Imagine our triangle sitting on the xy-plane. At each point (x, y) we evaluate the function R and plot the value at the appropriate height z above the point. As one might expect, the equilateral case (fig. 21.17) looks regular. The surface is a

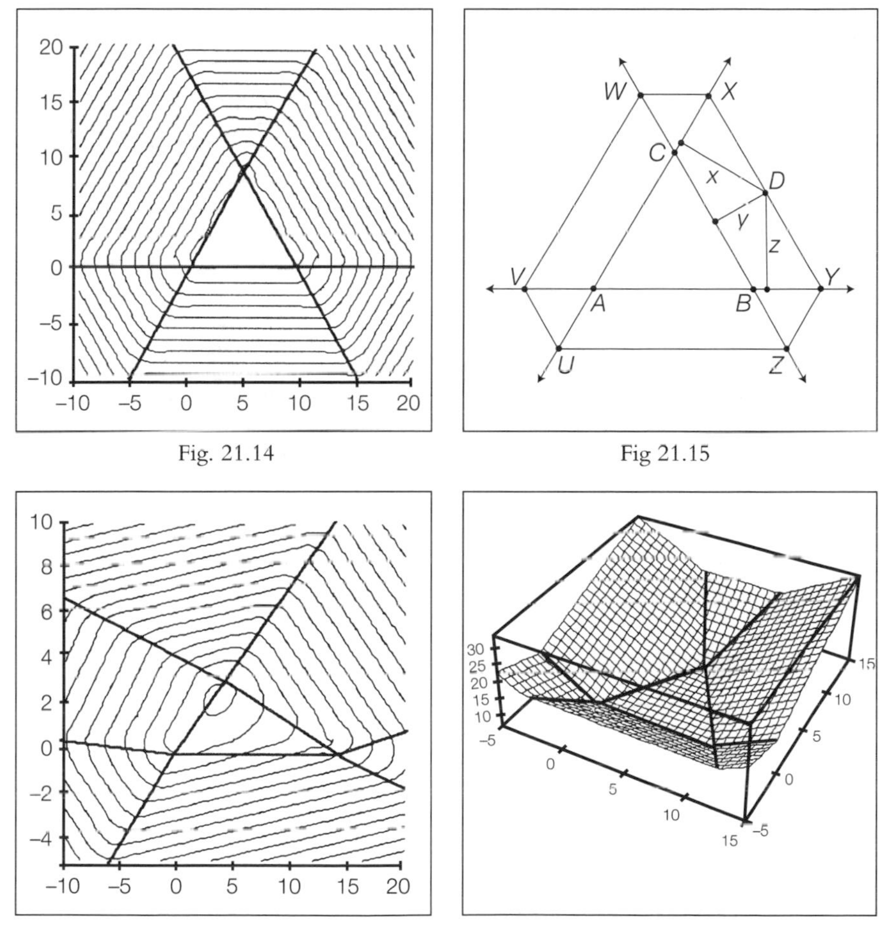

Fig. 21.14

Fig 21.15

Fig. 21.16

Fig 21.17

basin with planar walls. The points directly above the triangle form a level bottom for the basin. The contour lines may be interpreted as watermarks along the walls. And thinking about why the walls are *planes* instead of curved surfaces offers some insight into the original function. This is an appropriate place to make connections with students' studies of vectors.

One can use the same visualization technique to investigate the general case (fig. 21.18). Again the sides of the basin are planes, but the bottom of the basin—over the interior of the triangle—is no longer horizontal because the values of the function inside the triangle are not constant. This visualization helps us see why the watermarks will not be parallel to the sides of the triangle.

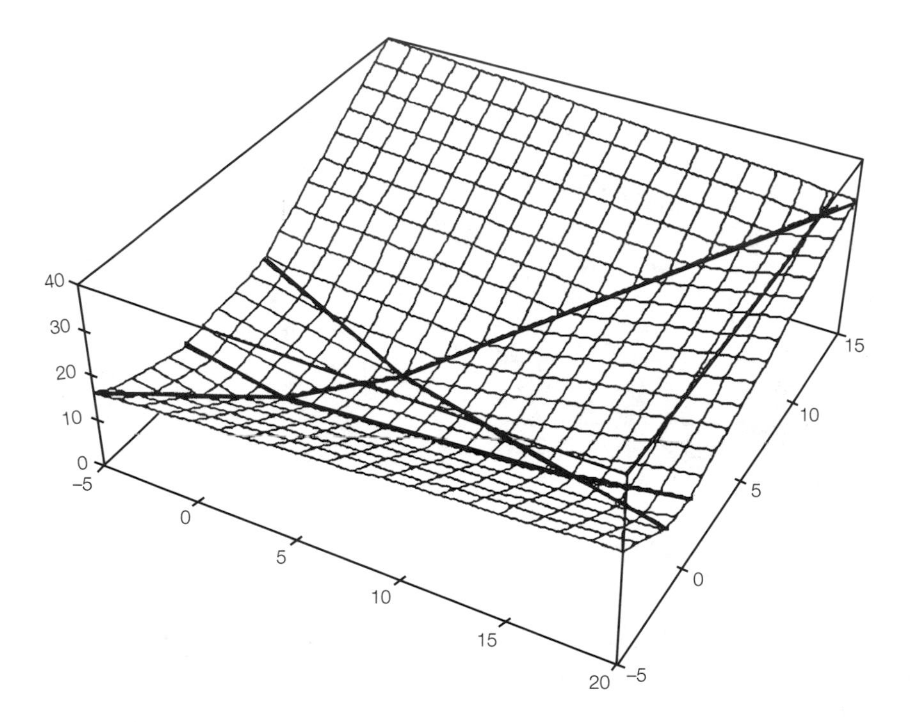

Fig. 21.18

4. The following activity offers a connection with physics: Imagine the vertices of a triangle as holes in a tabletop. Pass a string through each hole, and knot the three strings on top of the table. Below the table, hang three equal weights, one on each string. When the system falls to equilibrium, the forces on the knot must sum to 0, and (in a triangle where all angles are less than 120°) this can happen only if the three vectors along the strings make equal angles.

Connecting Discovering with Explaining

An interplay between experimental tools and an elementary but eclectic collection of methods for theoretical analysis makes it possible to connect geometry with other important and often traditional high school mathematics topics. More important than the particular topics visited, this approach allows students to experience parts of the method of mathematics that are often hidden from them: making definitions, using proof as an investigatory tool, posing problems, and modifying arguments.

In this investigation, as in mathematics itself, there is no separation between proof and experiment; for mathematicians, the search for explanations is a central technique in their research. Looking for a working justification for emerging and half-formulated facts that arise in the midst of experiments helps fine-tune conjectures and design new experiments.

To keep mathematics in the mathematics curriculum, curricula must emphasize this crucial feature of mathematical work. New technology allows the design of activities that put experimental power into the hands of students and teachers, but mathematics is more than the art of making data-driven discoveries. We certainly want more from students; and we believe that students are capable of more. Recent software is an indispensable tool in a student's laboratory for investigating, conjecturing, and verifying. But curricula must be crafted in a way that uses these tools to support and encourage a spirit of mathematical research by students. It is this use of technology, its potential for helping students discover *and* explain, that we find most exciting.

REFERENCES

Cabri Geometry II. Texas Instruments, Dallas, Tex., 1995.

Cabri: The Interactive Geometry Notebook. Brooks/Cole Publishing Co., Pacific Grove, Calif., 1992.

Education Development Center. *Connected Geometry: Building Roads to the Airport.* Dedham, Mass.: Janson Publications, in press.

Geometer's Sketchpad. Key Curriculum Press, Berkeley, Calif., 1990.

The Geometric superSupposer. Sunburst/Wings for Learning, Pleasantville, N.Y., 1992.

Klotz, Eugene. "Visualization in Geometry: A Case Study of a Multimedia Mathematics Education Project." In *Visualization in Teaching and Learning Mathematics,* edited by Walter Zimmermann and Steven Cunningham. Washington, D.C.: Mathematical Association of America, 1991.

Mathematica. Wolfram Research, Champaign, Ill., 1991.

Tangible Math: Geometry Inventor. Logal Software, Cambridge, Mass., 1994.

22

Forging Links with Projects in Mathematics

John W. McConnell

WHEN the mathematics teachers at Glenbrook South High School were challenged to broaden assessment through writing, they went beyond their commission to deal with critical issues of student involvement in learning, including the incorporation of technology into the classroom, the preparation for life skills in the next century, and the connection of textbook mathematics to the life of the community. The mathematics department devised what it has termed *mathematics writing projects.* These projects are based on issues and questions that justify the mathematics we expect our children to learn.

A project should furnish links to other academic disciplines or to the world in a way that supports later learning or uses previous learning. The central idea of the project should be one that is significant for mathematics, for another subject, or for general education. The project should require measurement and the acquisition of data, mathematical analysis, and a final paper or communication that represents the student's synthesis of the ideas embodied in the project. The project should advance the learning of mathematics appropriate at a particular stage in the curriculum. Projects in courses for younger students should take at least a day and increase in length and complexity as the students progress through the curriculum. More important than length is that there be evidence that the students are becoming *mathematized* (Romberg 1992) during their engagement in the project.

The two examples that follow illustrate the mathematics department's conception of projects for students in geometry and advanced algebra courses. These courses have had a long academic tradition that has separated the mathematical content from the world of the students; both these projects

The teachers who have prepared the projects cited in this article include Jack Adams, Jay Amberg, John Arko, Ann Brunner, Sandra Dawson, Amy Hackenberg, Marie Hill, Donna Hoffman, Kenneth Kerr, Tim Marks, Cathy Quigley, Kathy Schaumberger, Ann Wagner, and George Zerfass.

were designed to establish links between a unit in the respective course and the mathematics of business and commerce. The projects supply a context for substantial exploration with arithmetic computations linked to the concepts, formulas, and relationships studied in the classroom. Students formalize and generalize their explorations with papers that complete the projects.

A PROJECT FROM GEOMETRY

Geometry teachers are confident that their students learn how to use area and volume formulas in the situations found in textbooks or on tests. But can students select the correct formulas to use in a complex situation? Can they choose the appropriate units of measurement and make decisions that will optimize profit or minimize effort? Geometry teachers devised the project "You're the Estimator" to give an affirmative answer to those questions. Students played the role of an estimator for a concrete contractor in preparing a bid to install a concrete driveway and walkway for a house in our community.

The specifications given to the students included a diagram and the following information:

- The Glenview village code requires driveways to be six inches thick and sidewalks to be four inches thick.
- Delivered concrete costs the contractor $45.00 a cubic yard.
- Excavation and removal of debris costs $30.00 a cubic yard.
- Forms and their installation cost $0.50 a foot and must frame the entire perimeter of the project.
- *Puddling,* the labor of spreading the fresh concrete, costs $25.00 an hour. Puddling requires about five minutes for each square yard of surface.
- *Finishing,* or putting the final smooth surface on the concrete, costs the contractor $0.10 each square foot of surface.
- Once you compute the cost of the job, you need to add your profit margin to the cost. Any profit margin less than 15 percent is generally unacceptable; 15 percent is considered the minimum profit to stay in business.

One step in connecting the mathematics classroom to the life of the business world is teaching about that world. Since most of the high school students had never been in a bidding situation, the first task of the teachers was to educate their classes on the realities presented every day to the manager of a business. The homeowner wants the lowest price; the construction firm wants the largest profit. This reality is linked to the project by the grading of the final report:

> Write a job proposal so that your supervisor (the teacher) and the customer can understand how you arrived at your bid for this job. The project is worth 20 points:

- 5 points for correct grammar, spelling, and structure
- 5 points for writing style
- 8 points for a clear, step-by-step description of your mathematics
- *Bonus:* 2 points if you are among the five lowest bidders and at the profit margin
- *Penalty:* 2 points if you underbid and your company loses money
- No bonus or penalty if you make a profit but don't have a bid low enough to win the job

Some teachers had students work on the project as individuals. Others allowed students to work as teams. Figure 22.1 shows the front page submitted by a four-person team from a regular-level freshman class. They combined their names for the Geomar & Launic Contracting Company. The beginning of the letter to the homeowners, with spelling and phrasing as in the student work, is also shown. This paper had a good cover-sheet presentation, but the letter to homeowners reflected the youth of the students. The student letter continued for a full typed page. Students reported that the use of homeowners as an audience made it easier to start the writing process.

The teachers determined their own individual grading procedures, but they jointly determined what estimates would constitute the best solution for the project. The project provided practice with the volume and area formulas for solids and experience with the common measurement system used in the industry. It required that students decompose the driveway and sidewalk into fundamental objects for which they had volume formulas. There were many choices that produced good solutions. The report required the organization of the computational steps into a coherent argument.

A PROJECT FROM ADVANCED ALGEBRA

One of the most intense of our advanced-algebra projects is "A Matter of Some Interest," which is completed in two stages: data acquisition and problem solving. Classes gathered information about short-term investment opportunities from local banks and brokers and from the metropolitan newspapers. Teachers accumulated the options in a summary page, detailing features such as interest rate and compounding period and restrictions such as minimum value of investment and required length of time for the investment. Some information from the final bank-rate page is shown in table 22.1. The entire list included forty-five investment options from seven financial institutions. The data were used in class to compare the effectiveness of different rates before the students embarked on their assignment. The teachers used the data and graphing technology to contrast the rates and compounding periods. They exploited the different

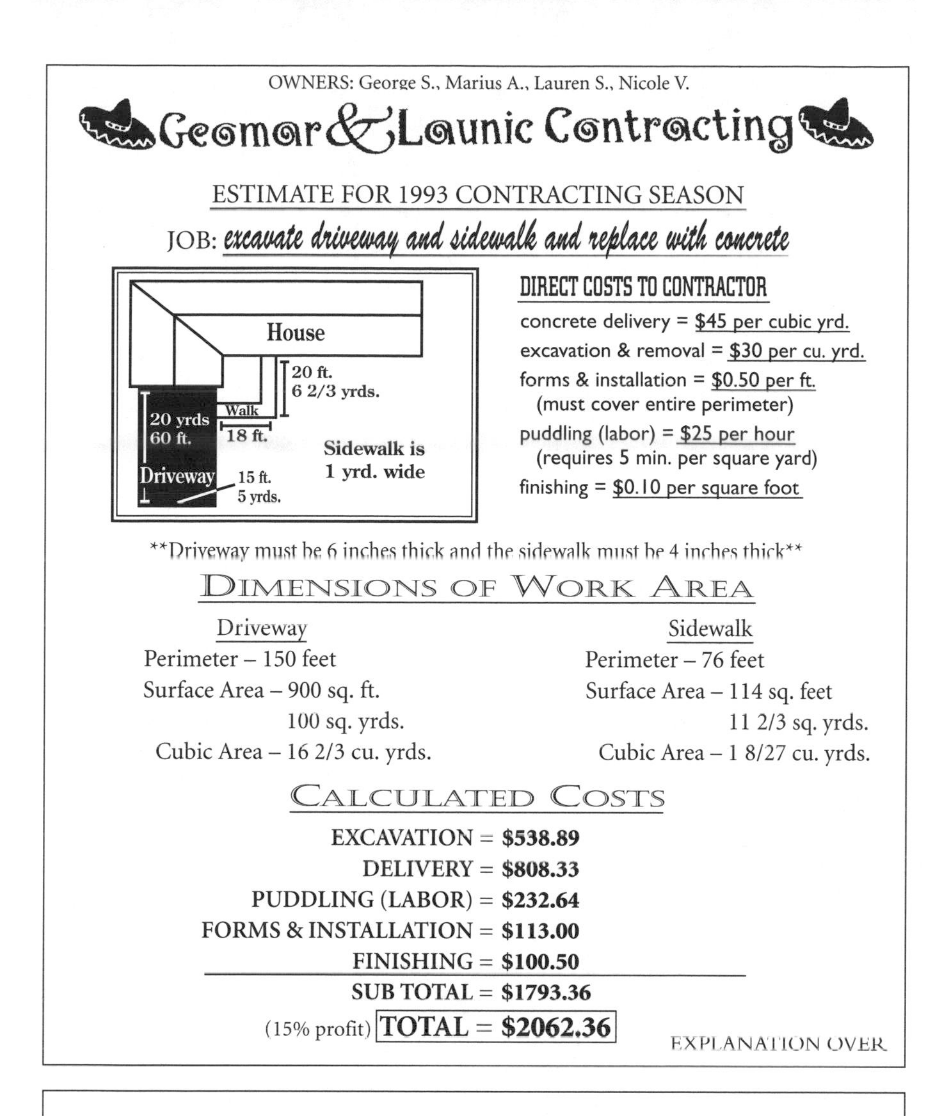

OWNERS: George S., Marius A., Lauren S., Nicole V.

Geomar & Launic Contracting

ESTIMATE FOR 1993 CONTRACTING SEASON

JOB: *excavate driveway and sidewalk and replace with concrete*

DIRECT COSTS TO CONTRACTOR

concrete delivery = $45 per cubic yrd.
excavation & removal = $30 per cu. yrd.
forms & installation = $0.50 per ft.
(must cover entire perimeter)
puddling (labor) = $25 per hour
(requires 5 min. per square yard)
finishing = $0.10 per square foot

Driveway must be 6 inches thick and the sidewalk must be 4 inches thick

DIMENSIONS OF WORK AREA

Driveway	Sidewalk
Perimeter – 150 feet	Perimeter – 76 feet
Surface Area – 900 sq. ft.	Surface Area – 114 sq. feet
100 sq. yrds.	11 2/3 sq. yrds.
Cubic Area – 16 2/3 cu. yrds.	Cubic Area – 1 8/27 cu. yrds.

CALCULATED COSTS

EXCAVATION = $538.89
DELIVERY = $808.33
PUDDLING (LABOR) = $232.64
FORMS & INSTALLATION = $113.00
FINISHING = $100.50

SUB TOTAL = $1793.36

(15% profit) **TOTAL = $2062.36**

EXPLANATION OVER

Dear Mr. and Mrs. Ichabod Peeps,

To ensure that you are not confused over how we at Geomar & Luanic Contracting arrived at our estimate, here is an explanation. We have also included our mathematic procedures if you doubt our conclusions.

The excavation and removal of the debree will cost us $30 per cubic yard. In the survey we calculated your driveway at 16 2/3 cubic yards (20 × 5 × 1/6). Therefore 16 2/3 cubic yards multiplied by 30 will give you the cost of excavation and removal of your driveway, $500. The sidewalk was calculated at 1 8/27 cubic yards ((1 × 5 × 1/9) + (1 × 6 2/3 × 1/9)). Using the same procedure as above, the cost of excavation and removal of your sidewalk will be $38.89....

Fig. 22.1

TABLE 22.1
Sample of Bank Rates Accumulated by Students in March 1993

Bank	Account Type	Annual Percent Rate	Minimum Deposit Time	Compounded	Minimum Deposit ($)
Glenview State Bank	passbook savings	2.70	—	quarterly	250
	money market	2.60	—	quarterly	2500
	6-month CD	3.05	6 months	simple interest	—
	12-month CD	3.35	12 months	annually	—
	2-year CD	4.10	2 years	quarterly	—
	3-year CD	4.50	3 years	quarterly	—
First Chicago	checking	2.00	—	daily	1000
	passbook savings	2.50	—	daily	100
	money market	2.55	—	daily	500
	6-month CD	2.75	6 months	daily	1000
	12-month CD	3.00	12 months	daily	1000
	2-year CD	3.35	2 years	daily	1000

annual, quarterly, and daily periods for compounding to introduce the constant e as the limit of $(1 + 1/n)^n$ where n increases without bound.

The second part of the project challenged students to pay debts that totaled $4375 and came due at different times over a three-year period. They were told that they had an initial capital of $4000. "In six months you will need $1000 to pay your car insurance premium; in one year, you would like to have $875 to buy a stereo system; in three years, you are planning a vacation to Hawaii that will cost $2500." Students needed to determine how to allocate their stake among the investment opportunities and justify their choices in a report. We encouraged students to use trial and error as well as reasoning skills. Students started with calculators, doing many more problems involving the exponential growth formula than they would in a conventional homework assignment. However, the huge number of choices that could be made in distributing the initial capital prohibited decision making purely by trial and error. After many computations most students realized that the schedule of debts favored certain investments, whereas the minimum-balance requirements eliminated others. Some teachers allowed their students to work with computer spreadsheets to build a model of the problem. Students wrote formulas and built a complex computational structure in a spreadsheet template. They were required to determine a good choice and schedule for investments and articulate their reasoning in a paper. Teachers graded students' reports on whether they took into account minimum required balances, stated the times for depositing and withdrawing money from accounts, were able to meet final expenses, and expressed their reasoning clearly and concisely.

Students did considerable arithmetic in their analysis of this problem. However, the project went beyond arithmetic. The mathematics included the exponential-growth formula, the use of a scale factor (1.0475) built from an interest rate (4.75%), and a sense of the effects of different interest rates and compounding periods. The work and analysis that students put into the paper accelerated the class discussion of the development of *e*. The summary paper required that the students articulate their reasoning using the formal language of functions and algorithms expressed in English.

The general quality of this project has been affected by the economy; it was easier to do when interest rates were high. Recent low interest rates reduced differences among investment opportunities and hence reduced the choices for students. This is indicated by the bonus points offered by teachers: a student can get 2 bonus points out of the 20-point scale by earning as little as $12 over expenses!

THE CURRICULAR AND INSTRUCTIONAL CONTEXT OF THE SCHOOL

These two examples represent a small sample of the variety of projects that have been developed by Glenbrook South High School teachers. A principal issue for the department has been the incorporation of projects into all courses and the implementation of critical projects into all sections of a course. Despite the pressures of an overloaded curriculum, external examinations, and a demanding community, the department has made major steps in establishing projects as a significant component of every Glenbrook South High School student's mathematics experience. A variety of institutional factors have accelerated the department's work with projects, many of which can be replicated in other schools and educational settings. These factors include making connections with other departments and schools, institutionalizing nontraditional assessment, and providing targeted in-service programs.

The department has been particularly active in using science as a foundation for projects. Many of its courses are built on textbooks that pay explicit attention to the foundation of great mathematical ideas in applications drawn from science. The Glenbrook faculty have worked with Argonne National Laboratories and Chicago State University in the development of computer-based courses organized about functions that are placed in the context of physics, biology, and commerce. Since 1991 the mathematics and science departments have shared a teacher who is charged with integrating essential ideas from physics into the advanced-algebra curricula. The collaboration with science is being replicated in English by pairing a mathematics teacher and an English teacher who have the specific mission of preparing a teamed course for freshmen. The use of one teacher in two departments or a teaching team that crosses departments can be accomplished without additional staffing resources.

An administrative fiat accelerated the department's use of projects within existing courses. In 1991, the Glenbrook superintendent decreed that all academic courses would give students two grades: a subject-specific grade and a writing grade. Fortunately, the mathematics department had been developing reading, writing, and communication strands in its curriculum in response to the National Council of Teachers of Mathematics (NCTM) view of "mathematics as communication" (NCTM 1989). When challenged to give a writing grade, the teachers asked a natural question: Write about what? Projects that connected to student life, to business and commerce, and to other school subjects seemed to offer the best topics for student writing.

Mathematics teachers were particularly aggressive in their development of in-service activities. They organized seminars that dealt with the impact of technology on curriculum and that enhanced their personal competencies with hardware and software. The typical seminar had themes such as *symbolic algebra systems* or *using the graphing calculator* that provided the organizing principle for the in-service activity. Participants shared the learning of technology, each mastering some features to present to the others. Every teacher prepared an activity for a class, implemented it with students, and then discussed the effects on instruction with the seminar group. The focus on technology freed teachers from extant curricula and enabled them to look forward to the mathematics instruction envisioned by national reports and recommendations. Teachers found that the most successful uses of technology emerged from data and experiments that linked to the real world—links that proved to be more important than the technology. Many of the projects that were developed in the technology seminars produced offspring that did not depend on any more technology than a scientific calculator. The chief components of the teacher-developed projects were the significance of the ideas and connections in the project, acquisition of data through research or measurement, mathematical analysis of the data, and a final paper or communication that synthesizes student learning. Technology may simplify or enhance any stage of the project, but it is not a necessity. As a result, when the board of education invested heavily in instructional technology, mathematics teachers designed a laboratory that took into account the low-tech activities of measurement, drawing, collaboration, and discussion as well as computers (McConnell 1992).

DIFFICULTIES WITH PROJECTS

Although Glenbrook South High School teachers have established a wide variety of projects within their mathematics courses, they do not have definitive answers to several questions. First, projects take time in and out of the classroom, so it is not possible to cover the content of textbooks. Some traditional topics must be deleted or deferred. What should they be? Second, what projects should be part of every student's school experience? States and school districts have no problem in creating lists of knowledge and skills that should be mastered by each student. If projects create encounters between students and great ideas, should all students experience the same projects? Or is doing any project sufficient? Third, how many projects should be in a course, since they take considerable energy on the part of teachers and students? Typically teachers do at least one project a quarter, but no course requires more than two in a quarter. Should there be one for every unit? More questions will appear as students progress through a series of courses that feature projects. Do projects really provide lasting knowledge? Do students develop skills that make learning new topics in mathematics easier?

We can answer one question in the affirmative: Do students and parents value projects? Yes. Parents are impressed with their children's work and report that they are happy the school is giving their children "the really important mathematics we didn't have." Students demonstrate their appreciation by the time they spend on projects; by their discourse in and out of class on the critical issues in a project; and by their summary papers, models, and speeches. These products arm teachers with a new way of measuring student learning and, because they are of high quality, translate into higher student grades.

FURTHER IDEAS FOR PROJECTS

Seesaw. Most algebra and advanced-algebra books ask students to solve problems involving balance on a seesaw. However, the park districts that serve our communities removed seesaws more than twenty-five years ago because of liability concerns. Therefore, our students do not have the same psychomotor sense of balancing different weights on a seesaw that the textbooks assume. Our teachers built a large seesaw that can be taken to a classroom or used in our mathematics laboratory. Using the law of the lever, the class determines how they might compute the weight of a student by balancing the student against a heavy, known weight. After the class demonstration, groups of three or four students go to the mathematics laboratory, where they are given a ten-pound dumbbell weight and asked to use the seesaw to find the weight of a mystery box of sand or bricks. Students must write out their procedures and solutions to the problem in individual papers.

Inverse Square Law. Some of the most important relationships in science and engineering can be expressed through the inverse square law. One of our projects in advanced algebra uses technology to develop the inverse square relationship for the intensity of light. We have connected a computer to a light sensor. As part of a classroom demonstration, students move a light source near the sensor, reporting to their classmates the distance from the sensor and the illuminance value reported on the computer screen. Students then plot the data points, composed of ordered pairs (distance, illuminance). They model the data with a graph of the form $I(d) = k/d^2$. Some teachers asked students to graph data by hand; others used spreadsheets.

Fast Ball. Each student tosses a tennis ball up in the air while another student times the flight of the ball. After measuring the height of the ball when it was released from the student's hand, each student develops a personal formula for the height of the ball as a function of time: $h(t) = (-1/2)gt^2 + v_0t + h_0$, where h_0 is the height at which the tennis ball was released and t is the time it took for the ball from release to hit the ground. If the ball has hit the ground at time t, $h(t)$ should be 0. Teachers supply the gravitational constant, g, so that students can solve the resulting equation for the initial velocity, v_0. Students then use graphing technology to plot the height of the ball for their toss as a function of time. We ask traditional questions about the toss: When did your ball reach its maximum height? How high was it one second after your toss? When will the ball be 300 meters in the air? The study of this most important quadratic function can progress more rapidly for students because the problem has been personalized for them. Students produce a short paper showing their personal parabola and telling how they answered the standard questions.

Gravity. We have attempted in several different experiments to encourage students to develop the gravitational constant. For example, we borrowed physics equipment so that our students could produce spark tapes that recorded the distance a dropped object fell during fractions of a second. Students modeled the resulting data with parabolas. In many instances, the scale factor a of the fitted parabola $y = ax^2$ gave an excellent estimate of $(1/2)g$. In other examples, the scale factor did not appear to be valid for any of the planets in our solar system. The activity created good data-gathering experiences for our students. Because the gravitational constant is one of the great ideas of physics, some teachers have extended the Fast Ball experiment. Some classes made videotapes of their ball tossing. The students then used a stop-frame video player to plot the height of their projectile as a function of video frame. (If you cover the video monitor screen with clear plastic wrap, you can trace the path and superimpose a grid with a marking pen.)

Miniature Golf. The physics of motion includes the analysis of bounces and caroms. The *UCSMP Geometry* text (Coxford et al. 1991) used in our regular geometry courses at the freshman and sophomore levels has an exciting lesson on "Miniature Golf and Billiards." We ask our geometry students to design a miniature-golf hole. They must write out an analysis of the best way of getting from the tee to the hole, using the methods of reflections demonstrated in the text.

Federalist Papers. Because students in our junior-level courses are studying or have studied American history, we ask them to replicate the Mosteller and Wallace statistical analyses that established authorship for some of the *Federalist Papers* of unknown authorship (Rubenstein et al.

1992). The social studies department has submitted three papers to the mathematics department related to issues the students will discuss in the American history course. One paper was authored by Alexander Hamilton, one by James Madison, and one was disputed. Students receive the documents without author designations. They mimic the Mosteller and Wallace technique of counting words and determining the frequency of certain prepositions and adverbs. Students work cooperatively in counting words and tallying data. This is a wonderful experience because they find that it is hard to count the words on a single page and that it is necessary to make decisions about what constitutes a word. Students submit individual papers in which they present their results and methods and justify their selection of Hamilton or Madison as the true author of their text. Students are also required to give an extremely brief (one or two sentences) summary of the content of their assigned *Federalist Paper,* a task that most consider more difficult than the mathematical computations.

Yearly Temperatures. In several courses students model temperature variations for a city over a period of a year. This project deepens the students' understanding of amplitude, phase shift, period, and the vertical shift of sinusoidal functions. We ask students to locate monthly average high or low temperatures for a temperate-zone city. They must use the library, on-line computer databases, or the computer software available in our laboratory as a source for their temperature tables. Students can use a variety of mathematics and statistical software for producing a scatterplot and fitting a model. They must assemble all this in a coherent paper that supplies a justification for the sinusoidal model and that gives information about the city they are discussing.

Population Growth. Students are assigned a state or country. They must use the library or other resources to find and plot the population over the last 100 years. Students in advanced algebra used linear and exponential models to make population predictions for the next 100 years. Some classes link sophisticated population models, such as the Verhulst (Olinck 1978; Peitgen and Richter 1986), with chaos theory, providing a bridge to readings such as *Jurassic Park* (Crichton 1990) and *Chaos* (Gleick 1987). Students use software such as Stella, Microsoft Excel, and Mathematica with special emphasis on the development of graphs, functions, and expressions that are the content of a contemporary advanced-algebra course. Again, students summarize their work in formal papers that are submitted for grading by both teachers.

Mathematics in Another Language. Some students who are new arrivals to the United States find the writing difficult. Our largest ESL (English as a second language) populations are Korean- and Spanish-speaking. Teachers asked students to prepare a short three-to-five minute oral presentation on a mathematical idea or famous mathematician. The students made the presentation in their native language, with another student serving as translator for the class.

Logic in Journalism. During their senior year, our regular-level students study forms of valid argument in mathematics (Perissini et al. 1992). Students select an argument from a newspaper or magazine editorial and parse the logic of the main idea using the symbolic logic of the mathematics course. The papers that result from this activity have a length that belies the intensity and difficulty of this task. The first problem students encounter is to find *any* logic in journalism. The second problem is to articulate the implied hypotheses and conditionals used by editorial writers, writing them in a form amenable to the algebra of logic. This activity requires substantial teacher-student discussion time in discerning arguments. Our graduates cite this project as one of the most memorable assignments in their high school careers and testify that it changed the way they read persuasive and political documents.

Sweete Shoppe. "Congratulations! You have just been promoted to head chef of the Sweete Shoppe, a fast-growing bakery in town. The last head chef was fired because he could not follow the recipes, which are given in matrix form." Students are given item-by-ingredient matrices. They collect prices on ingredients from local grocery stores, which are used in class to justify matrix multiplication. The writing project asks students to get two recipes from home, write a proposal to the Sweete Shoppe manager advocating the production of items as products, and use matrix multiplication to contrast the cost of ingredients from two suppliers.

Statistical Consultant. We use students as statistical consultants. Each year our student-activities office surveys the entire school to judge student involvement in, and satisfaction with, activities and athletics. Mathematics students analyze data from essential questions on the survey. They must determine what statistical techniques to use, decide whether their information is yielding something significant, and prepare a summary report for the director of student activities. This activity establishes an engaging link between textbook mathematics and the life of the school.

REFERENCES

Coxford, Arthur, Zalman Usiskin, and Daniel Hirschhorn. *UCSMP Geometry.* Glenview, Ill.: Scott, Foresman & Co., 1991.

Crichton, Michael. *Jurassic Park.* New York: Alfred A. Knopf, 1990.

Gleick, James. *Chaos: The Birth of a New Science.* New York: Viking Penguin, 1987.

Mathematica, Version 2.2, Wolfram Research, Champaign, Ill.

McConnell, John W. "Math Power." [In *The Electronic School,* a supplement of the] *American School Board Journal* 178 (September 1992): A36–A38.

Microsoft Excel, Version 4.0, Microsoft Corp., Redmond, Wash.

National Council of Teachers of Mathematics. *Curriculum and Evaluation Standards for School Mathematics.* Reston, Va.: The Council, 1989.

Olinck, Michael. *An Introduction to Mathematical Models in the Social and Life Sciences.* Reading, Mass.: Addison-Wesley Publishing Co., 1978.

Peitgen, Hans Otto, and P. H. Richter. *The Beauty of Fractals.* Berlin: Springer-Verlag, 1986.

Perissini, Anthony L., et al. *UCSMP Precalculus and Discrete Mathematics.* Glenview, Ill.: ScottForesman & Co., 1992.

Romberg, Thomas A. "Further Thoughts on the *Standards:* A Reaction to Apple." *Journal for Research in Mathematics Education* 23 (November 1992): 432–37.

Rubenstein, Rheta N., et al. *UCSMP Functions, Statistics, and Trigonometry.* Glenview, Ill.: ScottForesman & Co., 1992.

Stella, Version 2.10, High Performance Systems, Hanover, N.H.

23

Baseball Cards, Collecting, and Mathematics

Vincent P. Schielack, Jr.

THE importance of making mathematics more meaningful to students was recognized in the *Curriculum and Evaluation Standards for School Mathematics,* which calls for "opportunities to make connections so that students can use mathematics in their daily lives" (NCTM 1989, p. 32). Many hobbies enjoyed by students feature strong mathematical connections. Scale-model builders use the concepts of measurement, proportion, and similarity. Cross-stitch and needlework of all types use these same ideas, as well as the geometry of pattern reading and translation of a gridline pattern to cloth. Musicians use the fractions involved in musical intervals and key signatures. Computer enthusiasts require the logic of programming and the knowledge of algorithms and estimation procedures.

Sports and sports-card collecting are areas that interest a large number of students with widely varying mathematical backgrounds and abilities. We explore here two distinct types of connections between baseball cards and mathematics: player statistics and the mathematics of collecting. Statistical connections are accessible to elementary school students; connections involving collecting are more appropriate at the secondary school level.

BASEBALL-CARD STATISTICS

Baseball-card statistics typically can be categorized into two groups: those for pitchers and those for other players. In other words, a card has either pitching statistics or offensive statistics (fig. 23.1). Cards generally have career statistics for the pictured player, as well as statistics for one or more individual years. A legend for most common statistics is given in figure 23.2.

The calculated statistics create ample opportunities for students to see the concept of *function* in a familiar setting. For example, on nonpitchers' cards,

The author expresses his thanks to Dinah Chancellor and Jane F. Schielack for their help in the preparation of this article.

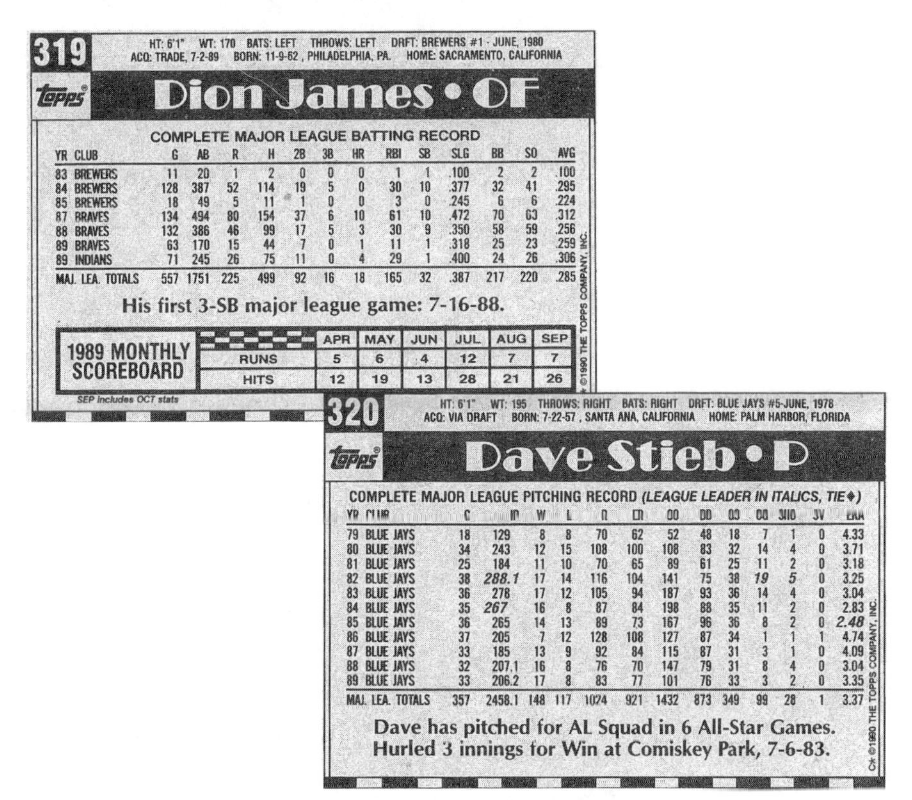

319 HT: 6'1" WT: 170 BATS: LEFT THROWS: LEFT DRFT: BREWERS #1 - JUNE, 1980
ACQ: TRADE, 7-2-89 BORN: 11-9-62 , PHILADELPHIA, PA. HOME: SACRAMENTO, CALIFORNIA

topps® **Dion James • OF**

COMPLETE MAJOR LEAGUE BATTING RECORD

YR	CLUB	G	AB	R	H	2B	3B	HR	RBI	SB	SLG	BB	SO	AVG
83	BREWERS	11	20	1	2	0	0	0	1	1	.100	2	2	.100
84	BREWERS	128	387	52	114	19	5	0	30	10	.377	32	41	.295
85	BREWERS	18	49	5	11	1	0	0	3	0	.245	6	6	.224
87	BRAVES	134	494	80	154	37	6	10	61	10	.472	70	63	.312
88	BRAVES	132	386	46	99	17	5	3	30	9	.350	58	59	.256
89	BRAVES	63	170	15	44	7	0	1	11	1	.318	25	23	.259
89	INDIANS	71	245	26	75	11	0	4	29	1	.400	24	26	.306
MAJ. LEA. TOTALS		557	1751	225	499	92	16	18	165	32	.387	217	220	.285

His first 3-SB major league game: 7-16-88.

1989 MONTHLY SCOREBOARD	APR	MAY	JUN	JUL	AUG	SEP
RUNS	5	6	4	12	7	7
HITS	12	19	13	28	21	26

SEP includes OCT stats

320 HT: 6'1" WT: 195 THROWS: RIGHT BATS: RIGHT DRFT: BLUE JAYS #5-JUNE, 1978
ACQ: VIA DRAFT BORN: 7-22-57 , SANTA ANA, CALIFORNIA HOME: PALM HARBOR, FLORIDA

topps® **Dave Stieb • P**

COMPLETE MAJOR LEAGUE PITCHING RECORD *(LEAGUE LEADER IN ITALICS, TIE♦)*

YR	CLUB	G	IP	W	L	R	ER	SO	BB	GS	CG	SHO	SV	ERA
79	BLUE JAYS	18	129	8	8	70	62	52	48	18	7	1	0	4.33
80	BLUE JAYS	34	243	12	15	108	100	108	83	32	14	4	0	3.71
81	BLUE JAYS	25	184	11	10	70	65	89	61	25	11	2	0	3.18
82	BLUE JAYS	38	*288.1*	17	14	116	104	141	75	38	*19*	*5*	0	3.25
83	BLUE JAYS	36	278	17	12	105	94	187	93	36	14	4	0	3.04
84	BLUE JAYS	35	*267*	16	8	87	84	198	88	35	11	2	0	2.83
85	BLUE JAYS	36	265	14	13	89	73	167	96	36	8	2	0	*2.48*
86	BLUE JAYS	37	205	7	12	128	108	127	87	34	1	1	1	4.74
87	BLUE JAYS	33	185	13	9	92	84	115	87	31	3	1	0	4.09
88	BLUE JAYS	32	207.1	16	8	76	70	147	79	31	8	4	0	3.04
89	BLUE JAYS	33	206.2	17	8	83	77	101	76	33	3	2	0	3.35
MAJ. LEA. TOTALS		357	2458.1	148	117	1024	921	1432	873	349	99	28	1	3.37

Dave has pitched for AL Squad in 6 All-Star Games.
Hurled 3 innings for Win at Comiskey Park, 7-6-83.

Reproduced with permission from the Topps Company, Inc.

Fig. 23.1

Baseball-Card Mathematics

Dinah Chancellor, 1992, Bryan, Tex., I.S.D.

What does it all mean? Read on!

G = Games
AB = At Bat
R = Runs
H = Hits
2B = Doubles (hits that allow the batter to advance to second base)
3B = Triples (hits that allow the batter to advance to third base)
HR = Home Runs
RBI = Runs Batted In
SB = Stolen Bases
SLG = Slugging %
BB = Bases on Balls
SO = Strikeouts
AVG = Batting Average
IP = Innings Pitched
W = Wins
L = Losses
ER = Earned Runs
GS = Games Started
CG = Complete Games
SHO = Shutouts
SV = Saves
ERA = Earned Run Average

Fig. 23.2

Batting Average is the number of hits divided by the number of times at bat. On pitchers' cards, *Earned Run Average* (ERA) represents the average number of earned runs the pitcher has allowed per nine innings. (An *earned run* is one that the team at bat earns with hits and walks, as opposed to a run that is allowed to score because of a fielding error.) Hence, the ERA can be calculated as the number of earned runs divided by the number of innings pitched, with that result multiplied by 9. The function concept is emphasized when students enter data and functions, such as those for batting average and ERA, into a spreadsheet program. Students can check their spreadsheet formulas and totals by comparing their calculated spreadsheet entries with those on the cards.

Some spreadsheet formulas for calculating card-back statistics can be entered only after calculating some intermediate statistics that are functions of those given. An example is *Slugging Percentage,* which represents the total number of bases attained divided by the number of times at bat. *Total Bases* is not on the card, but it can be calculated as follows:

4 times the number of home runs
+ 3 times the number of triples
+ 2 times the number of doubles
+ the number of singles

Singles are also not represented among the given statistics, but they can be calculated as follows:

number of hits
− number of home runs
− number of triples
− number of doubles

That is, all hits are singles except those that are not singles.

The following problems involving statistics can be explored by using spreadsheets, random-number generators, and other simulation techniques:

1. Currently a batter has 500 at-bats with a .290 batting average. If the batter expects 100 more at-bats over the remainder of the season, what batting average is necessary over the remaining 100 at-bats for the batter to become a .300-hitter for the season? (Questions similar to this can be asked regarding earned run average and slugging percentage.)
2. A *true .300-hitter* is one who theoretically would get a hit 30 percent of the time, given a very large number of at-bats. Of course, the actual number of at-bats in a season is a relatively small number, so that over the course of a season the batter's "luck" may result in a batting average higher or lower that .300. If the batter is a true .300-hitter and has 600 at-bats in a season, what is the probability that the batter will have a season batting average below .290? Below .280? (The theoretical probabilities that these lower-than-expected batting averages occur are about 28 percent and 13 percent, respectively, thus demonstrating that there is no "Law of Averages"!)

THE MATHEMATICS OF COLLECTING

Many of our students avidly pursue the collection of objects obtainable through a process that is essentially a blind draw. The collector acquires these objects either singly or in small groups, masked in some type of packaging so that the collector has no prior knowledge of which particular element in the set will be obtained from a given purchase. Examples include baseball cards and other types of cards depicting sports or music personalities, movie scenes, cartoon characters, and so on, as well as premium items, such as toys or stickers, packaged in a consumer product like cereal. (Although some baseball cards can be purchased as complete sets, many collectors still enjoy the "thrill of the hunt.") The following observations pertain equally to all set-collecting of the type described above, but for simplicity, we will henceforth refer to the collection of a set of baseball cards.

Many mathematical connections are involved in the process of collection. There are, of course, obvious elementary probability questions, such as: "For a set of ten items, if a collector already has seven, what is the probability that the next item acquired is one of the three needed to complete the set?" But the ones addressed here are more advanced problems whose solutions delimit the number of objects one would expect to have to obtain in order to complete a desired set, a number of sets, or a portion of a set. Some have solutions that involve probability, expected value, and infinite series (including geometric and harmonic series) in a natural form accessible to secondary school students.

All these problems, including those not lending themselves to closed-form solution, can be fruitfully attacked through Monte Carlo simulation techniques. The questions presented below can be simulated through a variety of means. Two- and six-card sets can be modeled with coins and dice respectively, and spinners and polyhedral dice can be used for other set sizes. The completion of larger sets can be modeled through computer simulation. These simulations can be used to make and test conjectures regarding solutions, to generalize results to sets of different sizes, and to initiate complete theoretical solutions when they exist.

We begin by making a few reasonable assumptions. We shall assume (1) that each card has an equally likely chance that it will be the next one we obtain and (2) that there are an infinite number of each card (that is, we assume that we draw cards with replacement). Obviously, there are never an infinite number of cards, but the number of each card produced in some sets is measured in millions, so for practical purposes this assumption models reality. Also, there are some older sets in which some cards are printed in greater quantities than others; the following questions do not apply to such sets (although the collection of such sets could also be simulated).

> **Question 1:** In a set of n cards, what is the expected number, N, of cards necessary to obtain each of the n cards at least once?

There are means of answering this question quickly in closed form using advanced statistical techniques (Feller 1967, p. 225). The solution can also be obtained using only elementary probability, expected value, and the sum S of the infinite series

$$\sum_{n=1}^{\infty} nx^{(n-1)} = 1 + 2x + 3x^2 + 4x^3 + \ldots,$$

where $|x| < 1$. This series looks almost like a geometric series, except that the coefficients of the terms increase arithmetically, so we shall call it a *quasi-geometric series.* We can find its sum, S, by using the familiar sum

$$G = \frac{1}{(1-x)}$$

of the geometric series

$$\sum_{n=1}^{\infty} x^{(n-1)} = 1 + x + x^2 + x^3 + \ldots,$$

where $|x| < 1$, as follows:

$$\begin{aligned} S &= 1 + 2x + 3x^2 + 4x^3 + \ldots \\ G &= 1 + x + x^2 + x^3 + \ldots \\ S - G &= x + 2x^2 + 3x^3 + \ldots = Sx \end{aligned}$$

So

$$(1-x)\cdot S = G = \frac{1}{(1-x)},$$

and

$$S = \frac{1}{(1-x)^2}.$$

With this formula for the sum in hand, we can now answer Question 1.

First, let us consider the special case of a set of three cards ($n = 3$) and designate the first card drawn as *a,* the second different card drawn as *b,* and the third different card drawn as *c.* The expected number of cards necessary to obtain *a* is 1. What is the expected number of cards needed to acquire *b* after getting *a?* We could obtain *b* on the first draw, and the probability of that happening is 2/3, since two of the three possible cards have not yet been drawn. Another possibility is drawing the sequence *ab,* which has probability (1/3)(2/3), so there is a (1/3)(2/3) probability of needing two cards to get *b* after getting *a.* A third possibility is drawing the sequence *aab,* which has probability $(1/3)^2(2/3)$. Or we could obtain the sequence *aaab,* which has probability $(1/3)^3(2/3)$. Continuing in this fashion, we find that the expected number of cards needed to acquire *b* after getting *a* is as follows:

$$1\cdot P(b) + 2\cdot P(ab) + 3\cdot P(aab) + 4\cdot P(aaab) + \ldots$$

$$= 1\cdot\frac{2}{3} + 2\cdot\frac{1}{3}\cdot\frac{2}{3} + 3\cdot\left(\frac{1}{3}\right)^2\cdot\frac{2}{3} + 4\cdot\left(\frac{1}{3}\right)^3\cdot\frac{2}{3} + \ldots$$

$$= \frac{2}{3}\left[1 + 2\cdot\frac{1}{3} + 3\cdot\left(\frac{1}{3}\right)^2 + 4\cdot\left(\frac{1}{3}\right)^3 + \ldots\right]$$

$$= \frac{2}{3}\left[\frac{1}{\left(1-\frac{1}{3}\right)^2}\right] = \frac{3}{2},$$

using the formula derived above for the sum of a quasi-geometric series.

Finally, we need to find the expected number of cards necessary to get *c* after getting *a* and *b.* The probability of getting the last remaining card in one draw is 1/3. Obtaining *c* after two draws results in the sequence _*c,* where the blank represents either card *a* or card *b;* the probability of this occurring is (2/3)(1/3). Extending this pattern, as above, we find that the expected number of cards to acquire *c* after getting *a* and *b* is as follows:

$$1 \cdot P(c) + 2 \cdot P(\_\,c) + 3 \cdot P(\_\,\_\,c) + 4 \cdot P(\_\,\_\,\_\,c) +$$

$$= 1 \cdot \frac{1}{3} + 2 \cdot \frac{2}{3} \cdot \frac{1}{3} + 3 \cdot \left(\frac{2}{3}\right)^2 \cdot \frac{1}{3} + 4 \cdot \left(\frac{2}{3}\right)^3 \cdot \frac{1}{3} + \ldots$$

$$= \frac{1}{3}\left[\frac{1}{\left(1 - \frac{2}{3}\right)^2}\right] = 3.$$

So N, the total expected number of cards to complete the three-card set, is

$$N = 1 + \frac{3}{2} + 3 = 5.5 \text{ cards.}$$

Using the model above, students can investigate other special cases for sets of four, five, six, … cards before attempting to answer the general case presented in Question 1. They will discover that in each case, the expected number of cards necessary to obtain the first card is 1. After obtaining the first card, the expected number of cards necessary to obtain the second (different) card is $n/(n-1)$. The expected number of cards necessary to get the third different card after getting the second is $n/(n-2)$. In general, the expected number of cards required to obtain the ith card after getting the $(i-1)$th card is

$$\frac{n}{n-(i-1)}.$$

Thus, the expected number of cards needed to obtain the last card is

$$\frac{n}{n-(n-1)} = n,$$

which agrees with intuition, since only $1/n$ of the cards available are the required last card.

Thus, for a set of n cards,

$$N = 1 + \frac{n}{(n-1)} + \frac{n}{(n-2)} + \ldots + \frac{n}{2} + n$$

$$= n\left(1 + \frac{1}{2} + \frac{1}{3} + \ldots + \frac{1}{n}\right)$$

$$= n(H_n),$$

where H_n represents the nth partial sum of the harmonic series. H_n can be approximated very accurately using $H_n \sim 0.57722 + \ln(n) + 1/2n$ (Graham, Knuth, and Patashnik 1989, p. 264), so that calculations can be performed without tedium. For example, in a twenty-card set, the expected number of cards to complete a set is $20(H_{20}) \sim 20(0.57722 + \ln(20) + 1/2(20) \sim 72$ cards.

Question 2. Sometimes cards are released in series. For example, a 600-card set may be released in two series of 300 cards each. How does this affect the expected number of cards needed to complete a set?

In a set of $2n$ cards in two series of n cards each, the expected number of cards needed to complete a set is twice the number for an n-card set, that is, $2n(H_n)$. If all cards were released together in a single series, the expected number would be $2n(H_{2n})$. Thus, the ratio of cards needed with two series versus one series is H_n/H_{2n}. For a 600-card set ($2n = 600$), this ratio is about 0.90, so that a two-series release would save the collector about 10 percent of the cards needed to complete a set.

Question 3. Some collectors desire cards of only certain players. For example, a collector may want the cards of only Nolan Ryan, Frank Thomas, Ken Griffey, Jr., Wade Boggs, and Roberto Alomar. What is the expected number of cards necessary to obtain these five cards?

The answer to this question depends heavily on the answer to Question 1. As far as the collector is concerned, it is as if the collector already has the other $n - 5$ cards. The next card (one of the five desired ones) will be expected to require $n/5$ cards, and the next four desired cards require $n/4$, $n/3$, $n/2$, and n cards, respectively. So we expect to need $n(H_5)$ cards to obtain the five desired ones. In general, if the collector seeks m particular cards from an n-card set, the expected number of cards required is $n(H_m)$.

Question 4. If the collector has obtained a random selection of C cards from a set of size n, how many different cards should be expected?

Obviously the answer is at most n. The probability of getting a particular card among the C cards is 1 minus the probability of not getting that card, or

$$1 - \left[\frac{(n-1)}{n}\right]^C.$$

Since each card has the same probability of being obtained, the expected value of the number of different cards is

$$n\left\{1 - \left[\frac{(n-1)}{n}\right]^C\right\}.$$

(Note that, as one might anticipate, this expected value increases and approaches n as C becomes larger and larger; that is, obtaining more and more cards results in an increasing number of *different* cards until the set is completed at some point.) For example, if a collector purchases a box of 540 cards from a set of 792, the number of different cards expected is about 392.

Question 5. What is the expected number of cards necessary to collect more than one complete set of cards?

An excellent simulation question deals with collecting more than one of the same set. Our formulas fail, since after the completion of the first set the remaining cards are not randomly distributed. One method of attack, however, might be to combine the formulas of Questions 1 and 4. For example, to complete two sets of ten cards each, we expect to need 29.3 cards to complete the first set, leaving 19.3 duplicates. When we abuse the formula of Question 4 to permit nonintegral values of *C,* these 19.3 cards yield an expected 8.7 different cards toward the second set, or 1.3 (one or two) more needed for completion of the second set. In a ten-card set, we expect to need 10 cards to obtain the last card, and 10/2 = 5 cards to obtain the next-to-last card. So the total expected number would be somewhere between 39.3 and 44.3 cards to complete the last set. A computer simulation of this problem situation could test whether this line of reasoning yields reasonable results and perhaps indicate how the theoretical model could be improved.

REFERENCES

Feller, William. *An Introduction to Probability Theory and Its Applications.* 3rd ed. New York: John Wiley & Sons, 1967.

Graham, Ronald L., Donald Knuth, and Oren Patashnik. *Concrete Mathematics: A Foundation for Computer Science.* Reading, Mass.: Addison-Wesley Publishing Co., 1989.

National Council of Teachers of Mathematics. *Curriculum and Evaluation Standards for School Mathematics.* Reston, Va.: The Council, 1989.

24

Experiencing Functional Relationships with a Viewing Tube

Melvin R. (Skip) Wilson
Barry E. Shealy

As is the case with all new concepts, the idea of relationship is best developed when it grows out of a child's concrete experiences with numerical relationships.

—E. R. Breslich, *Selected Topics in the Teaching of Mathematics,* Third Yearbook of the National Council of Teachers of Mathematics, 1928

Functional relations—that is, relations between quantities—will occur on every page of every book on mathematics unless we suppress them. We have been suppressing them.

—E. R. Hedrick, *Mathematics Teacher,* 1922

STUDENTS have a lifetime of experiences thinking informally about relationships among varying quantities. Teachers who successfully build on those experiences not only help their students appreciate the power of mathematics but also help them understand numerical relationships. In the activities described in this article, students have the opportunity to connect their informal knowledge with more formal mathematical ideas. Students also consider the relative value of, and connections among, several mathematical representations (e.g., verbal descriptions, graphs, tables, and formulas). Integrating graphing calculators and computer software makes the activities even more meaningful.

SEEING THROUGH A VIEWING TUBE: TWO EXPERIMENTS

Two closely related experiments engage students in measuring and recording data that model linear and inversely proportional relationships.

First, students explore the relationship between the distance one stands from a wall and the diameter or height of the circular area seen through a paper tube (fig. 24.1). For this experiment the length and diameter of the tube are held constant while the distance from the wall varies. In the second experiment the tube length varies while the distance from the wall remains constant.

Fig. 24.1

Making the Tubes

Perhaps the easiest materials for making the tubes are cardboard paper-towel or wrapping-paper tubes. Cutting the tubes to 25 cm works well for the first experiment. The second experiment requires tubes of several different lengths or a single telescoping tube. Telescoping tubes can be constructed by rolling up and fastening paper or cardstock to make two tubes and placing one tube inside the other.

Carrying Out the Experiments

Before students perform either experiment, they should record in their journals or some other appropriate place what they expect the relationship to be. Detailed predictions will make postexperiment reflection more meaningful.

Groups of three students work best for data collection: one student can be the "viewer," another the "measurer," and a third the "recorder." Vertically position and attach to the wall a meterstick centered about 1.5 meters above the floor. The viewer must look through the tube, holding it parallel to the floor, and direct the measurer to mark the extremes of the portion of the meterstick that the viewer can see through the tube (fig. 24.1). The measurer reads off the measurements to determine the diameter of the area that can be seen. The recorder writes the data in tabular form. Students should collect and record six to eight pairs of data for each experiment. For the first experiment, in which distance from the wall varies, ask students to measure the first height while the viewer stands about two meters from the wall, with subsequent distances at 0.5-meter increments. For the second experiment, in which tube length varies, the first tube length should be about twenty-five centimeters with increments of two to three centimeters. Table 24.1 shows actual student data for the second experiment.

TABLE 24.1
Student Data for Experiment 2

Length of Tube	Height of Viewing Area
25.0 cm	82.0 cm
27.5 cm	75.0 cm
30.0 cm	70.0 cm
32.5 cm	66.0 cm
35.0 cm	61.0 cm
37.5 cm	58.5 cm
40.0 cm	56.0 cm

Distance from wall = 5 meters
Tube diameter = 4.3 centimeters

Data can be analyzed using a TI-81 graphing calculator. After pressing [2nd] and [STAT], select the **DATA** menu (use the [▶] key), and then select **1:Edit** (press [ENTER],) Once the data are entered, make a scatterplot of the data by again pressing [2nd] [STAT], selecting the **DRAW** menu (use the [▶] key), then selecting **2:Scatter** by pressing [2] and [ENTER]. Remind students to be certain to select an appropriate range for the data before creating the scatterplot.

After the students have collected the data, they should discuss and write about the following questions in their groups:

1. As the distance (or tube length) increases, what happens to the size of the area seen on the wall? Be explicit.
2. Using your data, construct graphs illustrating the relationships. Do your tables and graphs tell you the same things? Is it easier to see and describe the relationships using a table or a graph? What are the advantages and disadvantages of each representation?
3. While keeping the length of the tube constant, how much of the wall do you think you could see if you stood right next to the

wall? How much could you see if you stood 20 meters away? While maintaining the same distance to the wall, how much of the wall would you see if you had a very short tube? An extremely long tube?

4. How are the relationships in the two experiments similar? Different?
5. Describe formulas that might model the relationships. Explain how you arrived at the formulas and why you think they are accurate. (*Note:* Graphing calculators significantly enhance this exploration; students can graph their formulas on top of the scatterplot.)
6. How did your prediction before the experiment compare to the results? Describe your reaction to the results. Were there any surprises?
7. Can you think of situations in the real world where these or other similar relationships might be relevant?

Whole-Class Discussion

After giving students time to reflect on their results, hold a whole-class discussion in which students report their results and discuss their interpretations. They may also share their results through formal written reports, oral reports, or poster displays. Because students base their understanding of these relationships on their experiences, they should be willing and excited to share their results.

A class discussion allows students to think more about their own conclusions while considering issues that may not have arisen in their groups. Ideally, the class will base its conclusions on class agreement following substantial debate. The teacher should ask students to consider connections among the different representations and to comment on which experiences (e.g., thinking about the experiment; conducting the experiment; recording data; constructing a table, graph, or formula) contributed most to helping them understand the relationships. Students should also discuss how each representation communicates different but important information. By not having single correct answers, these questions will challenge students to think critically and see mathematics as a subject in which "correct" depends on reason and circumstance, not simply on what is in the book or what the teacher says.

SEEING THROUGH A VIEWING TUBE: A GEOMETRIC ANALYSIS

These activities can be extended in a number of ways. For example, as students discuss the experiments, one or more of them may suggest a geometric interpretation of the situation. To explain the relationship between the distance to the wall and the height of the viewing area, a student may sketch a picture like that shown in figure 24.2. Considering similar triangles leads to explicit relationships among the varying quantities, thus extending the exploration described above (particularly Question 5).

Since $\triangle EBC$ is similar to $\triangle EDF$,

$$\frac{\text{diameter of area viewed}}{\text{tube diameter}} = \frac{\text{distance to wall}}{\text{tube length}}.$$

Thus, when one varies the distance to the wall (call this variable x) and holds the length of the tube constant, the relationship can be modeled with a linear function $y = k_1x$, where y is the diameter of the area viewed through the tube and k_1 is the quotient of the tube diameter and the tube length. Similarly, when the length of the tube (call this quantity x) is varied, a rational function $y = k_2/x$ models the relationship (k_2 is the distance to the wall multiplied by the tube diameter).

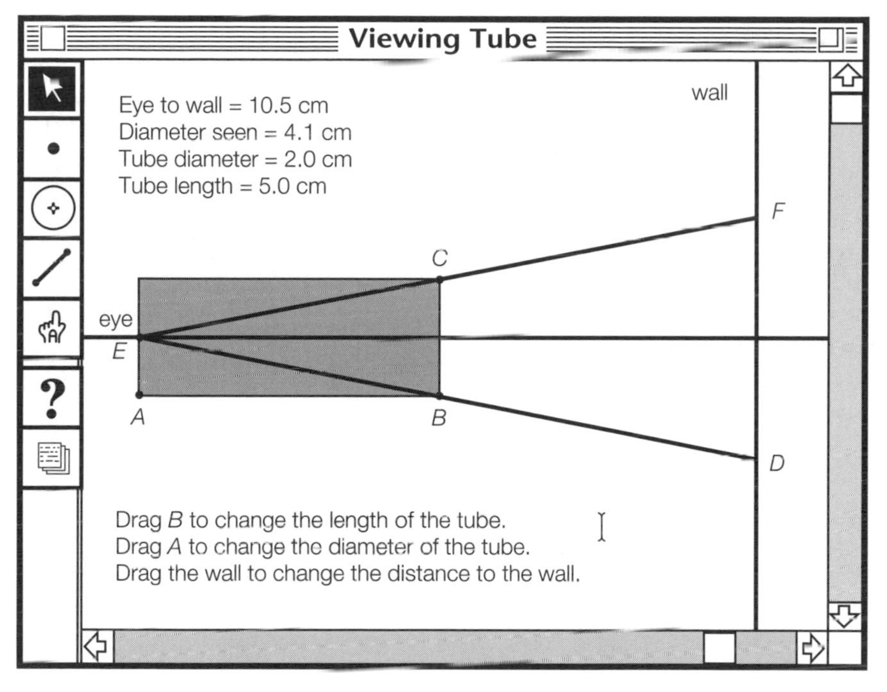

Fig. 24.2. A geometric interpretation

The teacher may model this interpretation or, better yet, allow students to explore it using such computer software as the Geometer's Sketchpad or Geometry Inventor. (Fig. 24.2 was created using the Geometer's Sketchpad by drawing a horizontal line and three lines perpendicular to it [the wall and the front and the back of the tube]. Line AB was constructed parallel to the horizontal line and then line EB was constructed. These two lines were then reflected about the horizontal line. We cleaned up the sketch by constructing segments [e.g., $\overline{AB}$] on the lines and hiding all the lines except the first horizontal line and the wall. Measurements and text complete the sketch.) With computer simulation, students can easily vary a third quantity: the tube's diameter. This reveals another linear relationship.

Students can also explore functions of more than one variable by considering the effects of allowing tube length, distance from the wall, and tube diameter to vary together. Such explorations can be significantly enhanced with the aid of computer technology. Spreadsheets or the built-in table-producing capability of the software mentioned above can help students manage and analyze data in these more complex situations.

CONCLUSION

The activities in this article offer opportunities for teachers to emphasize connections between formal mathematical ideas and students' informal experiences. These activities also draw attention to connections among mathematical representations as well as between algebra and geometry. Because the experiments require few materials, they are relatively easy to set up. Furthermore, students actually enjoy collecting the data. The most active discussion usually centers on whether both relationships are linear. The reflection and arguments that arise out of these activities lead to positive learning experiences and an enjoyable time for students. The experiential basis strengthens students' understanding of the function concept and gives them more flexibility in using this concept. Such strength and flexibility are particularly important, since function is a unifying idea (i.e., a great "connector") throughout school mathematics.

REFERENCES

Geometer's Sketchpad. Key Curriculum Press, Berkeley, Calif., 1992.

Geometry Inventor. Wings for Learning, Scotts Valley, Calif., 1992.

25

Breathing Life into Mathematics

Kristine Malia Johnson
Carolyn Leigh Litynski

MANY students have come to view mathematics as a static body of knowledge divorced from their world of experience. It is no wonder that few students think of mathematics as a dynamic field when fewer than 1 percent of all the mathematical concepts they learn were discovered after the eighteenth century (Steen 1978, p. 3). Establishing connections between the students' study of mathematics and the work of current mathematicians can challenge those misconceptions as well as "breathe life" into the mathematics classroom (Swetz 1984, p. 54).

Hoping to "breathe life" into our twelfth-grade class, we created a weeklong precalculus unit with the following two goals: (1) to demonstrate to students that mathematics is a topical, dynamic field; and (2) to engage students in thinking about, discussing, and exploring current mathematics.

The springboard for our lessons was an article by Richard Preston (1992), "Mountains of Pi," that appeared in the *New Yorker.* The 25-page article describes the Chudnovsky brothers, David and Gregory, and their search for order in the digits of pi. Preston's article focuses on rigorous mathematics, and at the same time it paints a colorful picture of two present-day mathematicians.

In Gregory's New York City apartment, the brothers Chudnovsky have built their own supercomputer, named *m-zero,* from mail-order parts. Their homemade computer is as powerful as a $30 million Cray computer but cost only $70 000 to build. The computer takes up Gregory's entire apartment. Circuit boards and stacks of paper output are everywhere, and more than twenty-five house fans blow air onto m-zero to keep it cool. The Chudnovskys never turn m-zero off for fear it will not turn back on again. Gregory and David believe that their homemade computer is the best machine on which to do their research, and consequently they do all their work out of Gregory's apartment. The Chudnovskys have had much success with m-zero. As of 1987, they held the world record for generating the most digits of pi, over 2 billion.

As the Preston article progresses, the mathematics becomes the prominent subject, with topics such as rational numbers, a history of pi, infinity, the Leibniz approximation for pi, transcendental numbers, and a proof of randomness. Thus Preston's article yields both traditional mathematics topics and a colorful context in which to learn the mathematics. Below we outline the lessons we created and include descriptions of how the lessons progressed.

LESSON 1: THE CHUDNOVSKY BROTHERS

- Read the first two pages of "The Mountains of Pi."
- Three-minute journal entry: "What does it mean to *do mathematics?*"
- Discuss journal entries and the Chudnovsky brothers.

We started the unit by reading aloud the first two pages of the article. Student responses in their journals included "Math is exercising preset laws ... and committing to memory those laws" and "Mathematics is a series of steps of a process.... When all parts of the process are complete, one should arrive at a predetermined answer." After hearing such responses, we were convinced that presenting mathematics as current discovery was a necessary experience for our students.

LESSON 2: RATIONAL AND IRRATIONAL NUMBERS

- Discuss reactions to the first twelve pages of the article.
- Three-minute journal entry: "What would you ask the Chudnovskys if you met them?"
- Discuss the history of irrational numbers (Preston 1992, p. 53).
- Formally define *irrational* and *rational* numbers.
- Explain how to write rational numbers as repeating decimals and the inability to do so with irrationals, such as pi.
- Present the formula for the sum of an infinite geometric series to show that 3/10 + 3/100 + 3/1000 + ... sums to 1/3.

We first explored the repeating-decimals and finite-sums-of-infinite-series properties of rational numbers by writing

$$0.333\ldots = \frac{3}{10} + \frac{3}{100} + \frac{3}{1000} + \ldots$$

on the board. (Students were in pairs at this point, with one graphing calculator to a pair, and they stayed in pairs for all the lessons.) The students found the fifth partial sum of this infinite series on their calculators and verified that it was close to 1/3. Next they summed the first six terms of the series and saw that adding one more term gave a number even closer to 1/3. We claimed that the sum of the entire infinite series is exactly 1/3, and noted that only rational numbers have this property: they can be exactly represented as the sum of an infinite series.

"Why can't we do that with pi?" John asked. He went to the board and attempted it:

$$3.1415\ldots = 3 + \frac{1}{10} + \frac{4}{100} + \frac{1}{1000} + \frac{5}{10\,000} + \ldots$$

"But what are you going to put next, John? There isn't a nice pattern like in 1/3. And how are you going to write those unpredictable numbers in sigma form?" queried Susan. John's unsuccessful attempt solidified the difference between rationals and irrationals for many of the students.

"But back to those sums for 1/3; how do we know they add up exactly to 1/3? Maybe they will be just very close to 1/3," commented Jan. The class decided that one way to verify the result was to test it in an existing formula, so we introduced the formula for the sum of an infinite geometric series:

$$S_n = \frac{a_0}{1-r}$$

After defining the variables in the formula, the students replaced the variables with the needed values, and the sum equaled 1/3.

"Well, fine, all that proves is that this formula you showed us works, but how do we know that the formula is valid?" asked Cheryl. As the bell was ringing, we had time only to respond that we would try to derive the formula tomorrow.

LESSON 3: FINITE SUMS OF INFINITE SERIES

- Derive the sum-of-infinite-geometric-series formula.
- Continue the discussion of rational versus irrational numbers.
- Introduce the Leibniz series as a representation of pi.
- Begin a group worksheet on estimating pi using finite sums of the Leibniz series.

We started class with the following derivation:

Let S be the sum of an infinite geometric series with a starting term of a_o and a constant ratio of r:

$$S = a_o + a_o r + a_o r^2 + a_o r^3 + \ldots \qquad (1)$$

Multiply both sides by r:

$$rS = ra_o + ra_or + ra_or^2 + ra_or^3 + \dots \qquad (2)$$

Subtract the second equation from the first:

$$S - rS = a_o$$

Simplify:

$$S = \frac{a_o}{1 - r}$$

The class eagerly followed the derivation, and after seeing the derivation, more students were convinced of the formula's validity.

We then asked the students to describe the Leibniz series, which was discussed in the previous night's reading assignment. A student wrote

$$\frac{\pi}{4} = \frac{1}{1} - \frac{1}{3} + \frac{1}{5} - \frac{1}{7} + \frac{1}{9} - \dots$$

on the board. Then the students tried to write the series in sigma notation. One representation they found was

$$\pi = 4\sum_{i=1}^{\infty} \frac{1}{2i - 1}(-1)^{i+1}.$$

Many comments and questions arose, including the following: "If we can write it in sigma notation, then why can't we just use our sum formula and get an actual number for pi? And how can you say pi isn't a rational number? Something is wrong somewhere!" This started a discussion of rational approximations for pi and the difference between geometric series and nongeometric series with predictable patterns, such as the Leibniz series.

The students next completed a worksheet to see how well the Leibniz series approximates pi. The worksheet, with sample student responses included, is shown in figure 25.1. The students arrived at their individual responses by first discussing each question collectively.

One outcome of the worksheet discussion was that students were making connections among the three ways to represent the Leibniz series: numerically (i.e., partial sums calculated in Question 1), symbolically (i.e., the expanded series), and graphically. Tom noted, "This is sort of making sense. The values we got for Question 1 are subsums of the entire series. The Leibniz formula seems to be saying that if we could sum the entire series, it would be exactly pi, and so subsums of the series are only approximations for pi, but pretty good approximations in my opinion. Since we alternately add and subtract values, it makes sense that the answers to Question 1 have a pattern of +/– errors. And the graph should show outputs that always hop above and below pi, and it does!"

Leibniz Series Worksheet

Solved for pi, the Leibniz series is

$$\pi = 4\left[\frac{1}{1} - \frac{1}{3} + \frac{1}{5} - \frac{1}{7} + \frac{1}{9} - \dots\right].$$

1. Calculate estimates for pi from 1 through 10 terms, to 5-digit accuracy.

1 term	*4.0000*	2 terms	*2.6667*
3 terms	*3.4667*	4 terms	*2.8953*
5 terms	*3.3397*	6 terms	*2.9760*
7 terms	*3.2837*	8 terms	*3.0171*
9 terms	*3.2524*	10 terms	*3.0415*

2. Why are approximations calculated from the even number of terms always less than pi?

   If you sum any finite amount, the sum will always be an estimate because you need all the infinite terms to get pi. An even number of finite sums always ends with subtraction, so the error will be on the minus side.

3. What can you conclude about approximations of pi using an odd number of terms?

   They will be off on the plus side of pi.

4. On the same set of axes:

   (*a*) Plot the ten data points from #1 on your calculator's plotter, with "term number" plotted on the horizontal axis and "pi approximations" plotted on the vertical axis.

   (*b*) Plot the line $y = \pi$.

   The following printout is based on a student's HP48G graphing calculator screen:

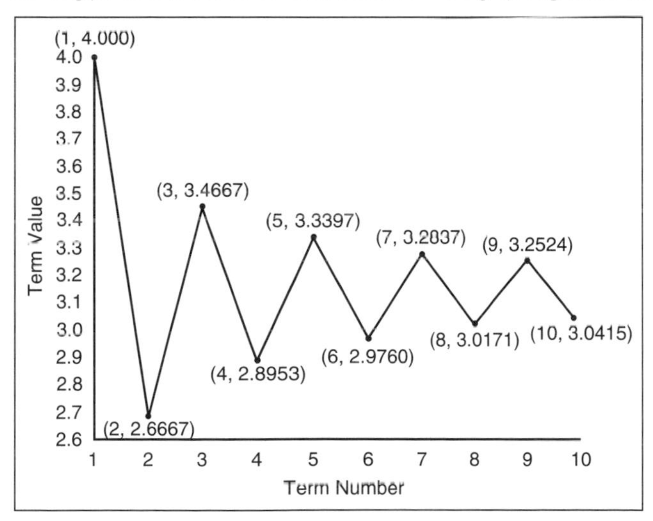

5. Do the data look like what you'd expect? Why or why not?

   Yes, the plots above pi and the plots below pi fit the plus/minus error idea of odd and then even sums.

6. Zoom in on the line $y = \pi$. Will this height ever be reached? Why or why not?

   Yes, at least it will look like it will reach $y = \pi$ because of that pixel problem with the calculator. At some point the estimate of π will be off by so little that on the calculator it will plot it on the $y = \pi$ line.

7. What does Preston mean when he writes, "The sum converges to $\pi/4$"?

   The sum gets closer and closer, and if you could get to all the infinite terms, it finally equals $\pi/4$.

8. Would you say that this series is converging quickly? Why?

   No, it is not converging quickly, because we added the first 10 terms and got an estimate for pi that is off by only about 1/10. To get exactly pi, we would have to add all the billions of other terms in the infinite series. So, to fix an error of only about 1/10,you need billions of more terms, which means that the sums must be moving in small steps to pi.

Fig. 25.1. Worksheet with student responses (shown in italic)

LESSON 4: MORE INFINITE SERIES

- Continue with the Leibniz worksheet.
- Discuss limits and summing infinite series.
- Present other infinite series that have finite sums.
- Discuss differences between theoretical and practical mathematics.

We finished the Leibniz worksheet, but there were still many questions. Lee commented, "I still don't see how an infinite list of sums, like that Leibniz series or that 1/3 series, can add up to anything but infinity, because even if it is a very small amount that you're adding, if you add an infinite amount of them, the sum becomes huge." We responded to this by holding up a sheet of typing paper that was drawn into boxes (fig. 25.2) and had a student start cutting on the lines.

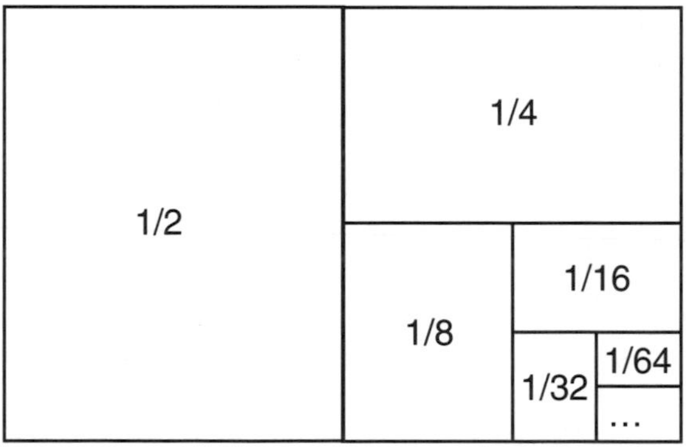

1/2
1/4
1/8
1/16
1/32
1/64
...

Fig. 25.2

"Susan cuts this paper in half, takes the remaining half and cuts that in half, and takes that remainder and cuts it in half, and keeps doing this forever. If we take all the cut pieces and add them together, how many whole sheets of typing paper will we have?"

"Well, of course you'll have one sheet, but be practical! Susan can't do that forever."

"True, we need to differentiate between theoretical and practical mathematics. For now let's just think in theoretical terms. If we *could* do this forever, then we would be adding an infinite number of pieces, like Lee observed. But we started with one sheet of paper, so all those infinite pieces must add to one piece of paper, not an infinite amount of paper." Next we wrote

$$1 = \frac{1}{2} + \frac{1}{4} + \frac{1}{8} + \frac{1}{16} + \frac{1}{32} + \frac{1}{64} + \ldots$$

on the board and made the connection between this series and all the cut pieces of paper. We used the geometric formula to verify that our intuition agreed with the mathematical theory:

$$S = \frac{\frac{1}{2}}{1 - \frac{1}{2}} = 1$$

Suddenly the students had many questions about infinity and about practical versus theoretical mathematics. We spent the remainder of class discussing how theoretical mathematics, though it sometimes cannot be actualized in the real world, can still be useful in the practical world. A student cited such an example from the article, recalling that in order for the Chudnovskys to theorize about the digits in pi, they needed to build a computer powerful enough to produce billions of digits.

LESSON 5: INFINITY IN THE PHYSICAL WORLD AND CONCLUSIONS

- Discuss Preston's description of infinite precision (1992, p. 42).
- Discuss: "Can infinity exist in the physical world?"
- Read and discuss "Pi in the Sky" (Waters 1990).
- Write for three minutes in journals: "What does it mean to *do mathematics?*"

We concluded this last lesson as we started the first, by discussing students' journal entries about what it means to do mathematics. After only one week, their definition of "doing mathematics" had broadened. Some entries included the following:

- "Math is a process that combines formulas, ...rules, ... and your own ingenuity to solve homework problems, household problems, and problems of universal import."
- "Math is not simply numbers and formulas; it takes imagination to develop 'new' math and to be able to use formulas to solve problems."
- "Math and doing math are the linking of the mind to the physical world."

This unit exceeded even our optimistic goals. We ended the week feeling satisfied that our students had learned traditional topics in a more relevant context: the Chudnovskys' search for order in the digits of pi. In addition, students learned about, and participated in, the vitality of mathematics. Even weeks after the unit ended, students found new patterns in pi.

REFERENCES

Preston, Richard. "The Mountains of Pi." *New Yorker,* 2 March 1992, pp. 36–67.

Steen, Lynn Arthur, ed. *Mathematics Today: Twelve Informal Essays.* New York: Springer-Verlag, 1978.

Swetz, Frank J. "Seeking Relevance? Try the History of Mathematics." *Mathematics Teacher* 77 (January 1984): 54–63.

Waters, Tom. "Pi in the Sky." *Discover* 11 (January 1990): 67.

FOR FURTHER READING

Jones, Phillip S. "The History of Mathematics as a Teaching Tool." In *Historical Topics for the Mathematical Classroom,* Thirty-first Yearbook of the National Council of Teachers of Mathematics, pp. 1–17. Washington, D.C.: The Council, 1969.

Lang, Serge. *The Beauty of Doing Mathematics: Three Public Dialogues.* New York: Springer-Verlag, 1985.

26

Students' Reasoning and Mathematical Connections in the Japanese Classroom

Keiko Ito-Hino

THE goal of emphasizing mathematical connections in the classroom is based on the premise that it is the student who will play the major role in making the connections. But what connections do students make naturally as they learn mathematics? And how can teachers capitalize on those connections in teaching mathematics? This article describes the reasoning exhibited by students during a series of eighth-grade algebra lessons in Japan. It comprises six lessons on simultaneous equations presented as three consecutive episodes. For each lesson, the intent, the problems used, and a brief summary of classroom activities are described, and students' reasoning in the classroom is illustrated along with three overlapping themes: (*a*) students used an informal approach and built on previously learned concepts and procedures when attempting to solve problems; (*b*) students acquired insight into mathematical ideas based on their own interpretation; and (*c*) students continued to seek algorithmic procedures. On the basis of these observations, suggestions for creating positive learning environments are made.

TEACHING SIMULTANEOUS LINEAR EQUATIONS IN JAPAN

Japanese students begin to learn algebra when they enter lower secondary school (grade 7). During the first year, they study positive and negative numbers, algebraic expressions using letters, equations with one variable, and direct proportion. Simultaneous linear equations, which are

The author acknowledges the useful reactions and suggestions of the editors and Professors Tatsuro Miwa, Jerry P. Becker, and Peggy A. House in completing this manuscript.

taught in grade 8, are included as an expansion of equations with one variable (Takakura and Murata 1990; Seki et al. 1990). Algebraic procedures are emphasized, although graphic representations of equations are treated in depth in another chapter on linear functions. Graphing calculators are not used at this time, even though the course of study refers to the use of calculators for simplifying computations.

In elementary and lower secondary schools in Japan, it is not uncommon for a whole class period to be devoted to one problem. The problem is presented, solved, and connected to specific mathematical content under the following general classroom organization (Becker et al. 1990, p. 15):

- Students' rising and bowing
- Reviewing previous day's problems or introducing a problem-solving topic (5 minutes)
- Understanding the topic (5 minutes)
- Problem solving by students, working individually, in pairs, or in small groups (20 to 25 minutes)
- Comparing and discussing (students write out proposed solutions on chalkboards) (10 minutes)
- Summarizing by the teacher (5 minutes)
- Assigning exercises (only 2–4 problems, to be done outside class)
- Sounding of a soft gong, indicating the end of class
- Students' rising and bowing

Teachers are encouraged to integrate the mathematical concept with the students' problem-solving abilities. Discussions between the teacher and the students and among the students themselves are considered important activities.

Episode 1: Introducing Linear Equations with Two Variables

Day 1 *Intent:* Explore an equation with two variables in a concrete situation. Share different solutions presented by the students.

The teacher showed students two twelve-sided dice. On a chalkboard, the teacher wrote the following problem:

> Die *A* and die *B* are twelve-sided dice. Suppose that you roll both dice at the same time. In what cases will the sum of "two times *A*" plus "*B*"equal fifteen? (Odaka and Okamoto 1982)

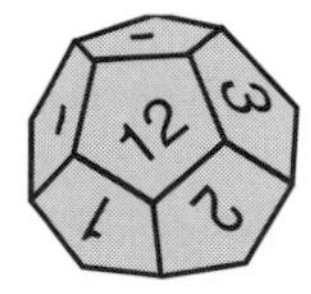

After students posed questions about the problem, they were asked to think about the problem on their own, to discuss it with their peers, and to solve it in their notebooks. The teacher walked around to observe what students were writing in their notebooks. Students who finished solving the problem were encouraged to solve it in another way. About twenty minutes later, some students volunteered to present their solutions on the chalkboard. The teacher offered some comments about the solutions and asked if any students had questions. Whenever students showed symbolic expressions of an equation with two variables, the teacher referred to this as "a linear equation with two variables."

Day 2 *Intent:* Examine the similarities and differences between the solutions of a linear equation with two variables and the solution of an equation with one variable. Understand that solutions of an equation with two variables can be given by ordered pairs (x, y).

After a brief summary of day 1, the teacher said, "Today I want all of you to solve the dice problem by using numerical expressions." Students thought about it and discussed it freely; presentations by several students then followed.

After the presentations, the teacher asked all the students, "How is the answer written?" Students discussed with the teacher and their peers the most appropriate way to express the answer. In relation to this, students were encouraged to examine similarities and differences between the solutions of an equation with two variables and the solution of an equation with one variable.

Informal Approach for the Initial Problem

Teachers always enjoy seeing how students attempt to solve the very first problem in a new chapter. Since students have not been directed toward the mathematical content yet, they approach the problem freely using only their previous knowledge. For the dice problem on day 1, a student got the answer $A = 5$ and $B = 5$ simply by guessing. Some students drew a 12-by-12 matrix ($2A = \{2, 4, 6, ..., 24\}$; $B = \{1, 2, 3, ..., 12\}$) to try every case (see fig. 26.1). Another student developed a table as in figure 26.2, which shows his reasoning. Still other students used logic and expressed their reasoning in sentences: "Since two times A is an even number, B must be an odd number; also, since die B is twelve sided, B must be less than 12 and greater than 0. So, B must be either 1, 3, 5, 7, 9, or 11, and it is easy to do the rest."

Every solution made sense with regard to finding the answer, though some involved more sophisticated mathematics than others. Furthermore, the solutions showed that students made different connections between the problem situation and mathematics. The manner in which the teacher presented the problem also had an effect on the mathematical connections the students made. For example, a student commented after the lesson that "when you showed us two dice at the beginning of the lesson, I thought that we were going to learn probability."

A	1	2	3	4	5	6	7	8	9	10	11	12
2A	2	4	6	8	10	12	14	16	18	20	22	24
B												
1	3	5	7	9	11	13	15	17	19	21	23	25
2	4	6	8	10	12	14	16	18	20	22	24	26
3	5	7	9	11	13	15	17	19	21	23	25	27
4	6	8	10	12	14	16	18	20	22	24	26	28
5	7	9	11	13	15	17	19	21	23	25	27	29
6	8	10	12	14	16	18	20	22	24	26	28	30
7	9	11	13	15	17	19	21	23	25	27	29	31
8	10	12	14	16	18	20	22	24	26	28	30	32
9	11	13	15	17	19	21	23	25	27	29	31	33
10	12	14	16	18	20	22	24	26	28	30	32	34
11	13	15	17	19	21	23	25	27	29	31	33	35
12	14	16	18	20	22	24	26	28	30	32	34	36

Fig. 26.1

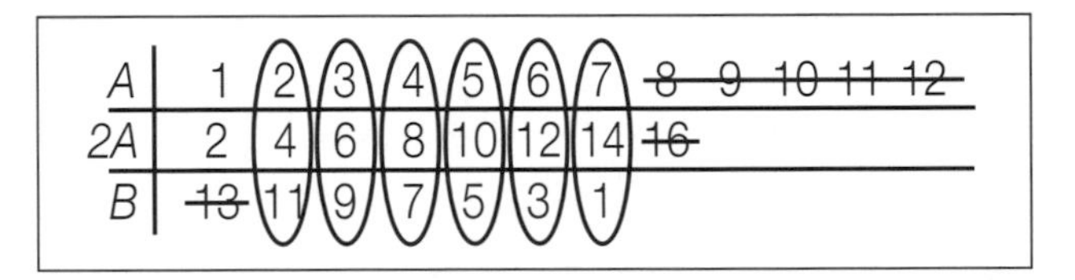

A	1	2	3	4	5	6	7	~~8~~	~~9~~	~~10~~	~~11~~	~~12~~
2A	2	4	6	8	10	12	14	~~16~~				
B	~~13~~	11	9	7	5	3	1					

Fig. 26.2

But enjoying different ways of solving a problem was not the sole purpose in using this approach; an important goal was to introduce linear equations with two variables and their solutions in a situation familiar to the students. In this regard, it was noted that fewer students used algebraic expressions than translated the problem directly into numerical expressions. Even when algebraic expressions were used, students preferred interpreting them as logic. One reason for this seemed to be that the students considered substitution to be less efficient and slower in computing many cases, as shown by the following exchanges:

Student 1: I made an expression $2A + B = 15$ and plugged numbers in for A and B.

Teacher: You substituted!

. . .

Student 2: I think that you can do this faster by changing $2A + B = 15$ to $B = 15 - 2A$. Answers for B can be figured out automatically by putting numbers into A.

In one classroom, no student presented substitution as a method of solution on day 1. On the second day, the students were told, "Let's solve this dice problem using the equation $2A + B = 15$." They suddenly quit exploring their own thoughts, and most began to devote themselves to inappropriate algorithms by manipulating symbols. It took some time for them to recognize substitution as an appropriate way to solve the equation. The interpretations students made were strictly procedural; they seemed to believe that in the mathematics classroom, algorithmic procedures should be given first priority in reaching the answer.

Making Connections between Solutions for Equations with One and Two Variables

Solution was not a new word for students because they used it when they studied equations with one variable during the previous year. However, the meaning needed to be expanded to apply to equations with two variables. The teacher anticipated students' confusion or lack of attention to this mathematical connection, so she intentionally discussed it with them on day 2. When the teacher asked for the solution to the equation $2A + B = 15$, students gave several responses, including the following: "There are many solutions for an equation with two variables, not just one," "The solution of $2A + B = 15$ cannot be determined because there is only one equation," and "It is impossible to get the solution because we can't solve this equation." Some students also said, "Solutions may always be obtained as 'sets.' " The word *set* in this context seemed to convey what students had in mind better than the word *pair.* Indeed, students showed a variety of ways to designate the sets of solutions:

$$A — \{2, 3, 4, 5, 6, 7\} \text{ and } B — \{11, 9, 7, 5, 3, 1\}$$

or

$$2 \leq A \leq 12 \text{ and } 1 \leq B \leq 11 \text{ (where } A \text{ and } B \text{ are whole numbers).}$$

Students considered "sets" to be two corresponding lists in which no further correspondence between members of the lists was recognized. Students pressed the word *sets* into service by trying to make a connection between solutions for equations with one variable and those with two variables.

Episode 2: Introducing Simultaneous Linear Equations

Day 3 *Intent:* Understand the concept of simultaneity in a concrete situation.

The teacher wrote the following problem (from Odaka and Okamoto [1982]) on the chalkboard:

> Die A and die B have twelve sides each. Suppose that you roll die A and die B at the same time. When do the dice satisfy the following two conditions:
>
> (i) The sum of "2 times A" plus "B" equals 15.
>
> (ii) "3 times A" minus "B" equals 5.

After asking several questions about the problem, students thought about it by themselves and with their peers. Following this, some students wrote their own solutions on the chalkboard. The teacher and students discussed the phrase *solution of two equations,* and the teacher introduced simultaneous linear equations:

$$2A + B = 15$$
$$3A - B = 5$$

Day 4 *Intent:* Investigate simultaneous linear equations in an abstract context. Understand that, in general, there are many (perhaps infinitely many) solutions of single equations with two variables, but there is normally only one solution for simultaneous linear equations with two variables.

The teacher gave students the problems below:

Obtain all the pairs (x, y) that satisfy each of the following conditions. (In these equations, x and y are integers.)

(1) $3x - y = -9$
(2) $3x + 2y = -18$
(3) $3x - y = -9$ and $3x + 2y = -18$

After checking the answers for cases (1) and (2), students were encouraged to think about the answer for case (3). Since many students got the answer (–4, –3) by computation, the teacher asked them, "Are there any other answers? Are you sure that (–4, –3) is the only solution?" Students then discussed their reasons. On the basis of what the students said, the teacher summarized that there is normally one solution of simultaneous linear equations with two variables.

Gaining Insight into a New Mathematical Idea: Simultaneity

The idea underlying simultaneous equations is the reconceptualization of one complex condition as the conjunction (intersection) of two independent and simpler conditions. To enhance the students' understanding of this fresh and important idea, the teacher suggested a problem context in which the idea was naturally embedded. As expected, students showed insights into the idea of finding the intersection even before they were asked to do so. On day 3, students developed two different ways of finding the intersection of the two conditions (i) and (ii): taking the common part and making the restriction. The following two solutions were developed by the students:

Solution A: Taking the common part of conditions (i) and (ii)

(i) $2A + B = 15$

A	2	3	4	5	6	7
B	11	9	7	5	3	1

(ii) $3A - B = 5$ or $B = 3A - 5$

A	1	2	3	4	5	6
B	-2	1	4	7	10	13

The case that appears in both tables is $A = 4$, $B = 7$. Therefore, $(A, B) = (4, 7)$ is the answer.

Solution B: Restricting the answer of condition (i)

The answers for condition (i) were obtained in Dice Problem 1. They are $(A, B) = \{(7, 1), (6, 3), (5, 5), (4, 7), (3, 9), (2, 11)\}$. Each pair must be checked to see if it satisfies condition (ii), $3A - B = 5$:

(7, 1)	→	$3 \cdot 7 - 1 = 21 - 1 = 20$	not satisfied
(6, 3)	→	$3 \cdot 6 - 3 = 18 - 3 = 15$	not satisfied
(5, 5)	→	$3 \cdot 5 - 5 = 15 - 5 = 10$	not satisfied
(4, 7)	→	$3 \cdot 4 - 7 = 12 - 7 = 5$	satisfied
(3, 9)	→	$3 \cdot 3 - 9 = 9 - 9 = 0$	not satisfied
(2, 11)	→	$3 \cdot 2 - 11 = 6 - 11 = -5$	not satisfied

Only the pair (4, 7) satisfies condition (ii); therefore, the answer is $(A, B) = (4, 7)$.

Problem familiarity may have led students to incorporate their knowledge from everyday life. Indeed, it is not rare to face decision-making situations with two different conditions by taking the common part or by checking whether one is okay for the other. For example, when determining the date for a meeting between two people, it is common for each person to list all available dates in a week. The meeting date is then selected from among the dates on both lists. Another possible way is for one person to make a list of her available dates and to ask the other person to check whether any of those dates are convenient.

However, we should not underestimate students' difficulty in the correspondence between the conjunction of conditions and its counterpart, the intersection of solutions. Indeed, when asked to think about the dice problem on day 3, the students posed questions about the meaning of "satisfying two conditions simultaneously." Discussion with the students revealed that they had several different interpretations of this wording. These interpretations included, "It means that both values, A and B, satisfy condition (i) at one time and condition (ii) at another time," "It means that both values, A and B, satisfy both conditions (i) and (ii)," and "It means that both values, A and B, satisfy condition (i)." They were confused about four different entities: value A, value B, condition (i), and condition (ii). Confusion also occurred when taking an intersection. There are six ordered pairs that satisfy condition (i); however, it was not easy for the students to deduce from this fact that one pair, (4, 7), satisfies both conditions (i) and (ii). They were confused because the value $A = 3$ appears for both conditions, as does the value $B = 1$. They did not recognize, for example, that the pairs (3, 9) for condition (i) and (3, 4) for condition (ii) do not produce a common solution for both conditions. Similarly they were confused because the value 5 appears as A and as B simultaneously for condition (i). The students needed additional effort to reinterpret the values within the problem context before they understood, for instance, that $A = 5$ and $B = 5$ do not satisfy both conditions.

Gaining Insight into Solutions for Simultaneous Linear Equations in an Abstract Context

Students are expected to decontextualize what they have learned in a concrete situation. On day 4, the students were asked to work with the three given cases in an abstract context. What seemed apparent was that in the abstract context, there were infinitely many solutions for cases (1) and (2), and in this circumstance, almost all students returned to using logic. However, it was not easy for them to determine conditions for their answers. Most students were overwhelmed by the fact that there were so many solutions, and they could not afford to explore certain conditions for such solutions. For example, a student showed the equation $3x - y = -9$ and said, "If I put $x = 1$, then $y = 12$; if I put $x = -54\,329\,345$, then $y = -1\,597\,389\,026$; but if I put $y = 100$, x is not an integer. Therefore, if x is an integer, there are infinitely many answers."

Considering these difficulties, it is reasonable that students paid attention to algebraic procedures for problem (3), which is the conjunction of problems (1) and (2). After they obtained the answer (–4, –3) for problem (3), the teacher intentionally asked the students if they were certain that the only answer was (–4, –3). One of the students said, "Since both (1) and (2) have many solutions, I think there are also many solutions for problem (3)." Another student commented, "The range of solutions of (3) should be smaller than the range of solutions of both (1) and (2), since (3) is the simultaneous equations of (1) and (2). I think that in this case, the range of (3) happened to be very small where only one solution exists." These insightful conjectures were refuted by other students, who stated, "The value of y for the solutions for (1) increases as x goes up. On the contrary, the value of y for (2) decreases as x goes up. It shows there is only one crossover between these x's." In one classroom, a graphic representation was also shown by a student to refute the conjectures.

Seeking Algorithmic Procedures

Even when students were successfully building on meaning, they still sought algorithmic procedures. Algorithmic procedures often overtook meaning, and this was illustrated dramatically at the end of day 3, when the following exchange occurred:

T: The answers (2, 11), (3, 9), (4, 7) … came from condition (i). What equation has these solutions?

Ss: $2A + B = 15$.

T: Okay. The answers (1, -2), (2, 1), (3, 4), (4, 7) … came from condition (ii). What equation has these solutions?

Ss: $3A - B = 5$.

T: Fine. Then what equations have the solution (4, 7)?

S1: There are too many of them. Can't figure it out.

T: Use today's equations.

S2: Both?

S3: I got it! I got it! I can write it.

The student walked to the chalkboard and wrote the following:

$$\begin{array}{r} 2A + B = 15 \\ \underline{+\ 3A - B = \ \ 5} \end{array}$$

S4: Oh! Now I got it!

S5: Great! It's perfect!

T: Wait! Wait …

The teacher expected to get the response of the simple expression of a simultaneous equation:

$$\begin{array}{r} 2A + B = 15 \\ 3A - B = \ \ 5 \end{array}$$

However, the student showed the algorithmic form for solving such equations. At that moment, it seemed clear that the student's focus was on the algorithmic aspect of the equation, and this focus seemed to pervade the classroom. The teacher was surprised that students were so inclined to see the algorithmic form. Some students even made comments in their notebooks that they understood the meaning of simultaneous equations very well because of this form. This observation illustrates students' habits as they make mathematical connections; they tend to make workable connections with regard to getting and expressing the answer. Presenting a real-world problem seemed to focus students on valid ways of finding the answer rather than on the meaning of simultaneity.

Episode 3: How to Solve Simultaneous Linear Equations

Day 5 *Intent:* Understand the substitution method by using diagrams.

The teacher gave students the following problems:

Solve each of the following pairs of simultaneous linear equations first without using algebraic expressions and then by using algebraic expressions.

(1) $x + 3y = 10$, $y = x + 2$ (2) $3x + y = 25$, $x + y = 18$

Students were first asked to solve the problems freely on their own and by talking with peers, then students presented their solutions for problem (1) without using algebraic expressions (i.e., by using diagrams). After sharing their ideas, students next presented solutions by using algebraic expressions. The teacher and students discussed how to make connections between the algebraic method of computation and using diagrams. The teacher introduced the phrase *substitution method* at the end of the lesson.

Day 6 *Intent:* Understand the steps of the addition-or-subtraction method by using diagrams.

Following day 5, students presented their solutions for problem (2) in the same manner as for problem (1). During the discussion about the connections between algebraic expressions and diagrams, the teacher asked the reason for the equivalence between $x + y = 18$ and $3x + 3y = 54$. At the end of the lesson, the teacher introduced the phrase *addition-or-subtraction method.*

Capitalizing on Previously Learned Concepts and Procedures

When solving problems (1) and (2) on day 5 and day 6, students showed methods of solution that were unexpected. Their methods were not always simple, and it required time to understand them. However, these methods reflected students' own insights into connecting a current problem to previously learned concepts and procedures. Two such methods developed by the students are shown below:

Solution of problem (2) from day 6 using a ratio method:

$$3x + y = 25 \quad \rightarrow \quad 54x + 18y = 450$$
$$x + y = 18 \quad \rightarrow \quad 25x + 25y = 450$$

Therefore, $54x + 18y = 25x + 25y$ and hence,

$$29x = 7y$$
$$x : y = 7 : 29.$$

(i) Take $x = 7$ and $y = 29$ and put them in $x + y = 18$. Then we get

$$x + y = 7 + 29 = 36.$$

To get 18 instead of 36, take $x = 7/2 = 3.5$ and $y = 29/2 = 14.5$.

(ii) Since the ratio of x to y is 7 to 29 ($x : y = 7 : 29$) and the whole is 18 ($x + y = 18$), we know that x is $7/(7 + 29)$ of 18 and y is $29/(7 + 29)$ of 18. Therefore:

$$x = \frac{7}{7+29} \cdot 18 = \frac{7}{36} \cdot 18 = 3.5$$

and

$$y = \frac{29}{7+29} \cdot 18 = \frac{29}{36} \cdot 18 = 14.5$$

The other method students developed was an area-model method. This method appeared in day 5 and day 6 and was used for problems (1) and (2).

Solution of problem (1) from day 5 using an area method (see fig. 26.3):

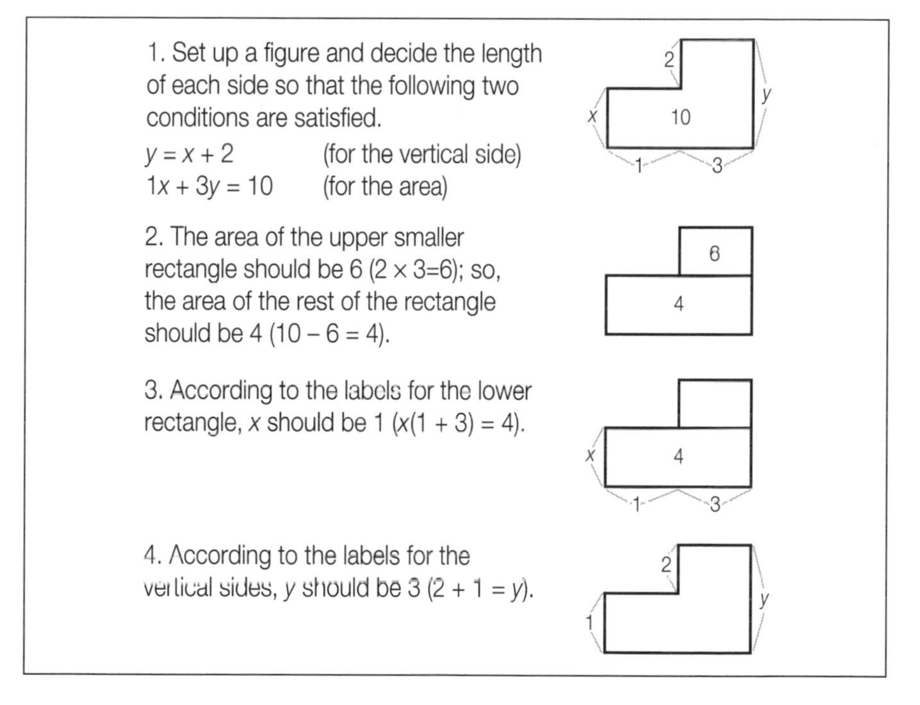

Fig. 26.3

Procedure over Concepts

It is interesting that the students seemed to prefer the addition-or-subtraction method to the substitution method. Some students used the addition-or-subtraction method both before and after its introduction in class on day 6. They used this method even when it would have been more convenient to use the substitution method. Their preference for the addition-or-subtraction method seems to relate to the issue of rote procedures versus genuine understanding of the method. The students' lack of understanding was observed not only in errors they committed but also in their questions such as, "What happens if the subtraction of an equation results in something like $2x + 3y = 10$?" and "How do we decide whether to add or to subtract?" which were written in their notebooks.

SUMMARY AND IMPLICATIONS

This article has described students' reasoning displayed during actual mathematics classroom lessons. Although the number of students observed (124) was small and the length of observation was short, important observations were made. First, the students brought into the classroom lessons a rich knowledge of previously learned concepts and procedures from everyday life. Such knowledge served as a base for giving meaning to a problem (even the initial problem), and it made possible the connection of the problem with the mathematics. Some of the students' problem-solving procedures appeared unexpectedly and showed unique connections. Second, students obtained and developed insight into new mathematical ideas based on such knowledge. Familiarity with the problem afforded students rich opportunities for developing insight. Students could interact with the knowledge in an effort to connect previously learned concepts and procedures to new ones. The connections were, however, easily colored by the students' focus on *getting the answer.* Third, it was difficult for the students to deal with an idea or a concept after they were informed about algorithmic procedures. Although algorithmic procedures often had little or no meaning for them, students persisted in using them because the procedures offered mechanistic tools and required little analytical thought. Students seemed to accept them without question.

The NCTM *Professional Standards for Teaching Mathematics* claims that

> [t]he importance of teachers' knowledge of how students learn mathematics cannot be minimized. Such knowledge provides direction for the kinds of learning environments that teachers of mathematics create, the tasks they select, and the discourse that they foster. (1991, p. 145)

Students' reasoning in the mathematics classroom suggests that we must pay more attention to the role of the problem contexts that foster mathematical connections made by students. The students described in this article could initiate their own connections using the given problems. Further, they could modify or even correct their connections in light of the problem contexts. That students were greatly affected by their familiarity with problem contexts implies that problems can be used to realize different connections. Teachers can also offer contrasts among different problem contexts (e.g., concrete and abstract) and encourage students to reflect on the different connections that they make. Filling a gap between informal and formal methods is also a method for developing the problem. The dice problem above can be solved using many informal methods, since it has finitely many answers; here the decontextualized problem on day 4 was important because it challenged those informal methods and gave a reason to develop formal methods for finding the one solution for simultaneous linear equations.

The difficulty of connecting concepts to procedures prompted Mack (1990) to write the following, after a careful analysis of students' reasoning in the teaching experiments:

> The results, however, do not suggest that the influence of rote procedures cannot be overcome, but a great deal of time and directed effort is needed to encourage students to draw on informal knowledge rather than use rote procedures. (p. 30)

One way to do this is to teach concepts prior to procedures and to stress making connections between concepts and procedures, as Mack and other researchers pointed out (e.g., Hiebert 1988). Also, it is possible to allow students to think about the rationale for the procedures that they want to employ by simply asking them "Why?" The students described in this article wanted to make reasonable decisions. For example, they wanted to know when to add and when to subtract when solving two simultaneous equations; they also wanted to know what needed to be done if the subtraction of equations results not in $2x = 10$ or $3y = 9$ but in something like $2x + 3y = 10$. Students can be encouraged to pursue their questions by asking themselves "Why does it not work?" or they can be asked for the idea common to both the substitution method and the addition-or-subtraction method. The thinking that results may offer an opportunity for students to connect their questions to the rationale for the algebraic method of solving systems of equations in general. Students can be encouraged to think "Why?" and to make mathematical connections by capitalizing on their own reasoning.

REFERENCES

Becker, Jerry P., Edward E. Silver, Mary Grace Kantowski, Kenneth J. Travers, and James W. Wilson. "Some Observations of Mathematics Teaching in Japanese Elementary and Junior High Schools." *Arithmetic Teacher* 38 (October 1990): 12–21.

Hiebert, James. "A Theory of Developing Competence with Written Mathematics Symbols." *Educational Studies in Mathematics* 19 (1988): 333–55.

Mack, Nancy K. "Learning Fractions with Understanding: Building on Informal Knowledge." *Journal for Research in Mathematics Education* 21 (January 1990): 16–32.

National Council of Teachers of Mathematics. *Professional Standards for Teaching Mathematics.* Reston, Va.: The Council, 1991.

Odaka, Toshio, and Koji Okamoto, eds. *Learning Tasks for Lower Secondary School Mathematics* (in Japanese). Tokyo: Toyo-Kan, 1982.

Seki, Setsuya, et al., eds. *Lower Secondary School Mathematics Two* (in Japanese). Tokyo: Dainihon-Tosho, 1990.

Takakura, Sho, and Yokuo Murata, eds. *Education in Japan: A Bilingual Text—Teaching Courses and Subjects.* Tsukuba, Japan: University of Tsukuba, 1990.